Palgrave Studies of Marketing in Emerging Economies

Series Editors
Robert E. Hinson
Durban University of Technology
University of Ghana Business School
Accra, Ghana

Ogechi Adeola
Pan-Atlantic University
Lagos Business School
Lagos, Nigeria

This book series focuses on contemporary themes in marketing and marketing management research in emerging markets and developing economies. Books in the series cover the BRICS (Brazil, Russia, India, China and South Africa), MINT (Mexico, Indonesia, Nigeria and Turkey), CIVETS (Colombia, Indonesia, Vietnam, Egypt, Turkey, and South Africa); EAGLE economies (those which are expected to lead growth in the next ten years, such as Brazil, China, India, Indonesia, South Korea, Mexico, Russia, Taiwan, and Turkey) and all other African countries (classified under developing countries), taking into consideration the demographic, socio-cultural and macro-economic factors influencing consumer choices in these markets. The series synthesizes key subject areas in marketing, discuss marketing issues, processes, procedures and strategies across communities, regions and continents, and also the way digital innovation is changing the business landscape in emerging economies.

Palgrave Studies of Marketing in Emerging Economies presents a unique opportunity to examine and discuss marketing strategy and its implications in emerging economies, thereby filling a gap in current marketing literature.

All chapter submissions to the series will undergo a double blind peer review and all book proposals will undergo a single blind peer review.

More information about this series at
http://www.palgrave.com/gp/series/16591

Ishmael Mensah
Kandappan Balasubramanian
Mohd Raziff Jamaluddin
Gina Alcoriza • Vanessa Gaffar
S. Mostafa Rasoolimanesh
Editors

Marketing Tourist Destinations in Emerging Economies

Towards Competitive and Sustainable Emerging Tourist Destinations

palgrave
macmillan

Editors
Ishmael Mensah
University of Cape Coast
Cape Coast, Ghana

Kandappan Balasubramanian
Taylor's University
Selangor, Malaysia

Mohd Raziff Jamaluddin
Universiti Teknologi MARA
Selangor, Malaysia

Gina Alcoriza
University of Santo Tomas–Legazpi
Manila, Philippines

Vanessa Gaffar
Universitas Pendidikan Indonesia
Bandung, Indonesia

S. Mostafa Rasoolimanesh
Taylor's University
Selangor, Malaysia

ISSN 2730-5554 ISSN 2730-5562 (electronic)
Palgrave Studies of Marketing in Emerging Economies
ISBN 978-3-030-83710-5 ISBN 978-3-030-83711-2 (eBook)
https://doi.org/10.1007/978-3-030-83711-2

This Palgrave Macmillan imprint is published by the registered company Springer Nature Switzerland AG.
The registered company address is: Gewerbestrasse 11, 6330 Cham, Switzerland

Preface

Many emerging economies are turning to tourism as the new economic generator and are not sparing efforts at developing their tourism potentials. This has heightened competition among tourist destinations. However, emerging economies also have to compete with the well-established destinations in North America, Australia, and Europe. The ability of emerging economies to compete favourably on the international tourism market is constrained by a number of crises including terrorism, political unrest, natural disasters, pandemics, and accidents, which tend to affect the image and attractiveness of such destinations. In addition, emerging economies suffer from institutional and structural deficiencies as well as marketing challenges occasioned by the emergence of new tourist destinations, disintermediation in the distribution system, inadequate ICT infrastructure, changes in consumer behaviour, and fragmentation of the tourism industry. Addressing these challenges calls for a well-coordinated management and marketing strategy.

The tourist destination is evolving, and new developments have implications for destination marketing in emerging economies. The debilitating effects of COVID-19 call for a new approach to destination marketing in emerging economies. Also, advances in ICT have given rise to new concepts such as digital marketing, collaborative marketing, social media marketing and green marketing, e-tourism, and virtual tourism (v-tourism), which has led to the emergence of the concept of smart destinations. The Internet offers destination marketing organizations (DMOs) a more cost-efficient option for reaching out to a global audience through social media and other digital platforms. Thus there is the need for a book that synthesizes and addresses all the issues confronting emerging economies in their destination marketing efforts. However, most of the texts on tourism destination marketing and management reflect destinations in more advanced countries. The body of knowledge on destination marketing in the context of emerging economies is patchy and scattered. This book seeks to bring all the destination marketing issues and concepts relating to emerging destinations together in a single book. Readers will understand all the nuances and contemporary issues and developments in the field of destination marketing as applied to emerging economies.

It is important to acknowledge the fact that emerging tourist destinations have unique characteristics and peculiar challenges which have implications for destination marketing. This book highlights the destination marketing challenges, best practices, and strategies relevant to emerging economies.

Cape Coast, Ghana Ishmael Mensah
Selangor, Malaysia Kandappan Balasubramanian
Selangor, Malaysia Mohd Raziff Jamaluddin
Manila, Philippines Gina Alcoriza
Bandung, Indonesia Vanessa Gaffar
Selangor, Malaysia S. Mostafa Rasoolimanesh

Acknowledgements

This book project would not have been accomplished without the support of some individuals and organizations. First, we wish to express our sincerest appreciation to the series editors Prof. Robert Ebo Hinson and Prof. Ogechi Adeola for believing in us and allowing us to contribute to the book series 'Palgrave Studies of Marketing in Emerging Economies'. We also wish to acknowledge all the authors and anonymous reviewers without whom this project would have been a failure. Our special thanks go to the Team at Springer Nature, including Rubina Infanta Rani, the Project Coordinator, for their immense support. Finally, we acknowledge Mr Emmanuel Agbaglo and Ms Eunice Sakyi Kyeremah for editorial assistance.

Contents

Notes on Contributors

Evelyn Addison-Akotoye is a lecturer in the Department of Hospitality and Tourism Management, University of Cape Coast. She holds a PhD in Hospitality Management from the same university. Her research interests cover issues related to nutrition literacy, food choice behaviour, menu management and service quality.

Gina Alcoriza is an assistant professor and the programme chairperson for the Tourism Management Programme at the University of Santo Tomas–Legazpi, Legazpi City, Philippines. She holds a doctorate in Business Administration from the University of Santo Tomas–Legazpi, Philippines.

Eunice Fay Amissah is a senior lecturer and the head of the Department of Hospitality and Tourism Management, University of Cape Coast. She holds a PhD in Hospitality Management with a specialisation in service quality management from the University of Cape Coast. Her research interests include workplace diversity, work-family conflict and service quality.

Maria Criselda G. Badilla is an associate professor at the University of the Philippines Asian Institute of Tourism. She is also a tourism market specialist and branding expert. Her research interests are in tourism marketing, destination branding and digital media. She has published textbooks on tourism, which are used in tertiary and senior high schools in the Philippines.

Mohd Azlan Abu Bakar is the Dean for Faculty of Management, PICOMS International University College. He has lectured at the same institution since 2018. He holds a professional degree from the Chartered Institute of Logistics and Transport (CILT), UK, and a Master of Science in Transport and Logistics. His research interests include supply chain management and logistics management.

Kandappan Balasubramanian is an associate professor with the School of Hospitality, Tourism and Events, Taylor's University, Malaysia, and an Apple Distinguished Educator. His research interests include Industrial Revolution 4.0, service innovation in hospitality 4.0, teaching innovation towards transformation in the educational and industry sector, marketing in the hospitality industry as well as in higher education.

Sarah Blankson-Stiles-Ocran is Assistant Lecturer in Hospitality Management at the Ho Technical University, Ghana. She holds an MPhil degree in Hospitality Management from the University of Cape Coast. Her research interests include employees' well-being, employees' work attitudes, intentions and behaviour, as well as food handling, recipe development and food tourism.

Reil G. Cruz is an associate professor at the University of the Philippines Asian Institute of Tourism, where he also serves as the Institute Secretary. He served as an international consultant in Human Resource Development, Marketing, and Livelihood Generation for the UN World Tourism Organization in East Timor, Sri Lanka, and North Korea from 2005 to 2006. His research interests include tourism planning and development, tourism impact and sustainability, ecotourism, and sustainable tourism.

Joy Sheelah B. Era is the Dean of the Institute of Tourism and Hotel Management, Far Eastern University, Manila. She holds a PhD in Philippine Studies and MA in Art Studies from the University of the Philippines-Diliman. Her other affiliations include Accreditor of the Philippine Association of Colleges and Universities Commission on

Accreditation (PACUCOA), and she has recently been assigned as a Regional Quality Assurance Team (RQUAT) of the Commission on Higher Education (CHED) for the tourism and hospitality programmes.

Mavis Adjoa Forson is an assistant lecturer at the Department of Tourism and Hospitality Management, Ho Technical University. She holds an MSc in Leisure, Tourism and Environment from Wageningen University in the Netherlands and an MA in Hospitality Management from the University of Cape Coast. Her research interests are tourism and livelihoods, sustainable tourism development, and food and beverage services.

Vanessa Gaffar is Associate Professor of Marketing in the Faculty of Economics and Business Education, Universitas Pendidikan Indonesia. She is a Vice Dean for Academic Affairs. She has been a visiting researcher at Coventry University, UK. Her research interests are in the field of tourism marketing, sports tourism, customer relationship management, social media, and entrepreneurship.

Emmanuel Gamor is a PhD student at the School of Hotel and Tourism Management, the Hong Kong Polytechnic University, and a faculty at the Department of Hospitality Management of the Koforidua Technical University, Ghana. He holds a Master of Philosophy in Hospitality Management from the University of Cape Coast. He is also a Certified Hospitality Educator (CHE) and a Certified Hotel Industry Analyst (CHIA) by the American Hotel and Lodging Association. His research interests include sustainable tourism, human resource management, tourism marketing, and tourist behaviour.

Wan Soraya Wan Abdul Ghani is a lecturer for the Undergraduate Event Management Degree Programme in the Faculty of Business and Management, Universiti Teknologi MARA (UiTM), Malaysia. She holds a master's degree in Hotel and Tourism, majoring in Event Management from the University of Queensland, Australia. Her research interests include event management, sporting events, event marketing, and tourism marketing.

Candra Hidayat is Assistant Professor of Hotel Management at the Trisakti School of Tourism, Indonesia. He received his Graduate Double Degree from Trisakti School of Tourism, Indonesia, and Burapha University, Thailand. He is pursuing a doctoral programme at the Trisakti School of Tourism, Indonesia.

Mohd Raziff Jamaluddin is a senior member of the Faculty of Hotel and Tourism Management in the biggest university in Malaysia, Universiti Teknologi MARA. He holds a PhD from Universiti Teknologi MARA. He also holds the position of Adjunct Faculty at Alagappa University (India) and Visiting Professor at PICOMS International University College, Malaysia. His research interests include destination branding, hospitality consumer behaviour, event management, and safety and security of tourism destinations.

Ishmael Mensah is Professor of Tourism and Hospitality Management and Ghana Director of the Confucius Institute at the University of Cape Coast. He is also an external expert at Hunan City University. He holds a PhD in Tourism from the University of Cape Coast as well as postgraduate certificates in Hospitality Administration and Event Planning and Tradeshow Management from Georgia State University, where he was a fellow under the Ghana Tourism Capacity Development Initiative. His research interests include service quality management, destination marketing, environmental management in hotels, special events management, and community-based tourism development.

Christopher Mensah is an associate professor at the Department of Hospitality and Tourism Management, Ho Technical University. He received his PhD in Tourism from the University of Cape Coast, Ghana. His research interests are in the field of sexual harassment in hospitality and tourism, customer satisfaction, hospitality human resource management, hospitality education and special events.

Benjamin Appiah Osei is a PhD graduate from the School of Hospitality, Tourism and Events, Taylor's University, Malaysia. His research interests include travel motivation and decision-making, application of advanced technologies in hospitality, social media and travel behaviour, and eco-citizenship behaviour.

S. Mostafa Rasoolimanesh is an associate professor at the School of Hospitality, Tourism and Events, at Taylor's University. Rasoolimanesh is an editorial board member of more than ten reputed tourism and hospitality journals including the *Journal of Hospitality and Tourism Technology*, *Journal of Hospitality and Tourism Management*, and *International Journal of Tourism Policy*. His research interests include sustainable tourism, heritage tourism, destination planning and management, and urban sustainability.

Agus Riyadi is Assistant Professor of Hotel Management at the Trisakti School of Tourism, Indonesia. He had his master's degree from Universiti Teknologi MARA, Malaysia, and is pursuing a PhD programme at the Universiti Teknologi MARA, Malaysia. He has 18 years of experience in the hotel industry and higher education in tourism, with expertise in marketing research.

Richmond Sakyi is an assistant lecturer in the Department of Hospitality and Tourism Management, Ho Technical University. He holds a Master of Philosophy in Tourism Management from the University of Cape Coast. His research interests include tourism development, ecotourism, hospitality and tourism marketing, and corporate social responsibility.

Abbreviations

AAA	African African-American
AEC	ASEAN Economic Community
AMPC	ASEAN Master Plan for Connectivity
AR	Augmented Reality
ASEAN	Association of Southeast Asian Nations
BAS	Branding-Advertising-Selling
BAS	Branding, Advertising, and Selling
BRI	One Belt One Road
CBBE	Customer-Based Brand Equity
CHSE	Cleanliness, Health, Safety, and Environmental Sustainability
COMCEC	Committee for Economic and Commercial Cooperation of the Organization of the Islamic Cooperation
COVID-19	Coronavirus Disease 2019
CRS	Computer Reservation Systems
CSFs	Critical Success Factors
DMO	Destination Marketing Organization
DOT	Philippines Department of Tourism
EDM	External Destination Marketing
ETC	European Travel Commission
EU	European Union
GDS	Global Distribution Systems
GMTI	Global Muslim Travel Index
GTA	Ghana Tourism Authority

HS09	Homecoming Scotland 2009
IATA	International Air Transport Association
ICT	Information and Communication Technologies
IDD	Internal Destination Development
IMF	International Monetary Fund
IMFITP	It's More Fun in the Philippines
IOC	International Olympic Committee
LADA	Langkawi Development Authority
MATTA	Malaysian Association of Tour and Travel Agents
MERS	Middle East Respiratory Syndrome
MICE	Meetings, Incentives, Conventions and Exhibitions
MME	Malaysia Major Events
MyCEB	Malaysia Convention and Exhibition Bureau
NAACP	National Association for the Advancement of Coloured People
NATO	North Atlantic Treaty Organisation
NTAs	National Tourism Authorities
NTO	National Tourism Organization
OECD	Organisation for Economic Co-operation and Development
OTA	Online Travel Agent
PANAFEST	Pan African Historical Theatre Festival
PATA	Pacific Asia Travel Association
POSE	Paid Media, Owned Media, Social Media, and Endorser
ROPAB	Representation of the People Amendment Bill
SARS	Severe Acute Respiratory Syndrome
SDGs	Sustainable Development Goals
SM	Social Media
SMEs	Small- and Medium-Sized Enterprises
SPBM	Strategic Place Brand Management
TAST	Transatlantic Slave Trade
TCL	Travel Career Ladder
TTCI	Travel and Tourism Competitiveness Index
UAE	United Arab Emirates
UNESCO	United Nations Education Scientific and Cultural Organization
UNWTO	United Nations World Tourism Organization
VAC	Visit ASEAN Campaign
VR	Virtual Reality
WEF	World Economic Forum

WHO	World Health Organization
WTO	World Tourism Organization
WTTC	World Travel and Tourism Council
YOR19	Year of Return 2019

List of Figures

List of Tables

Part I

Destination Attributes and Features

1

Nature and Scope of Destination Marketing in Emerging Economies

Ishmael Mensah, Mohd Raziff Jamaluddin, Vanessa Gaffar, Gina Alcoriza, Kandappan Balasubramanian, and S. Mostafa Rasoolimanesh

Introduction

Emerging economies, which are also known as emerging markets, are low- or middle-income economies with high economic growth potential. However, they are usually less stable and imperfect with regard to the efficiency and impartiality of their markets due to lapses in their

---

I. Mensah (✉)
University of Cape Coast, Cape Coast, Ghana
e-mail: ikmensah@ucc.edu.gh

M. R. Jamaluddin
Universiti Teknologi MARA, Selangor, Malaysia
e-mail: raziff@uitm.edu.my

V. Gaffar
Universitas Pendidikan Indonesia, Bandung, Indonesia
e-mail: vanessa@upi.edu

I. Mensah et al. (eds.), *Marketing Tourist Destinations in Emerging Economies*, Palgrave Studies of Marketing in Emerging Economies,
https://doi.org/10.1007/978-3-030-83711-2_1

institutional frameworks, which are commonly known as institutional voids (Meyer et al. 2016). They are countries that are transitioning from the developing phase to the developed phase. These economies have lots of potentials and have become increasingly important in the global economy in recent years. Emerging economies have also become very important as tourist destinations. Tourism growth over recent years has largely been driven by emerging economies (Kester & Croce, 2011). Emerging tourist destinations are becoming particularly attractive to tourists from the key generating markets in North America and Europe because they have comparative advantages in cultural, ecological and other alternative tourism products. In spite of this, emerging economies are far from achieving their full potentials, as they are confronted with many challenges, which hinder effective destination marketing. Because emerging destinations have to compete with the well-established tourist destinations and battle with the adverse impacts of crises such as the COVID-19 pandemic, emerging destinations have to adopt more innovative and strategic approaches to destination marketing.

For emerging destinations to achieve their full potentials, they must strive to be competitive in the global tourism market. This requires destination marketing organisations (DMOs) in emerging economies to develop and execute a well-thought-out marketing strategy. This introductory chapter tackles the foundational issues of emerging economies, emerging tourist destinations and destination marketing. The chapter focuses on issues such as the definition of concepts of destination marketing, tourist destination and emerging economies; the geographical scope of emerging economies; as well as features of emerging tourist destinations, destination marketing challenges of emerging tourist destinations,

---

G. Alcoriza
University of Santo Tomas–Legazpi, Manila, Philippines
e-mail: gina.alcoriza@ust-legazpi.edu.ph

K. Balasubramanian • S. M. Rasoolimanesh
Taylor's University, Selangor, Malaysia
e-mail: kandappan.balasubramanian@taylors.edu.my; mostafa.rasoolimanesh@taylors.edu.my

destination mix and destination marketing organisations in emerging economies. The chapter also sets the tone for the rest of the chapters, examining the implications of the issues discussed for destination marketing in emerging economies.

The Tourist Destination

The concept of tourist destination became popular in the post–World War II period when most countries started opening their borders for people and trade. The tourist destination has been defined as a system. According to Tinsley and Lynch (2001, p. 372), it is a "system containing several components such as attractions, accommodation, transport, and other services and infrastructure". Others have defined the tourist destination as a geographical area (Beirman, 2020; Buhalis, 2000). Buhalis (2000) describes a destination as a geographical region serving integrated services to tourists and composed of a combination of the tourism products or the places with distinct natural attractiveness and properties that may be appealing to the tourists. Thus, a tourist destination is a geographical space in which several tourism resources exist. These include attractions, infrastructure, equipment, service providers, other support sectors and administrative organisations whose integrated and coordinated activities provide tourists with the experience they expect from the destination (Fyall et al., 2019). The geographical elements of Leiper's (1995) tourism system model (Fig. 1.1) can be used to explain the scope of the tourist destination. The geographical elements include (a) the traveller-generating region, (b) the tourist destination region and (c) transit route region.

The traveller-generating region (or home region) refers to the place where the tourists come from. It is the generating market that stimulates and motivates travel. The transit route region includes both the short period of travel from the tourists' home region to the destination and other places on the way that the tourists may stop to visit. The tourist destination region is one of the most important elements in the whole tourism system. It is composed of suppliers who cater for the needs of the tourists, physical resources which are crucial for attracting tourists, the

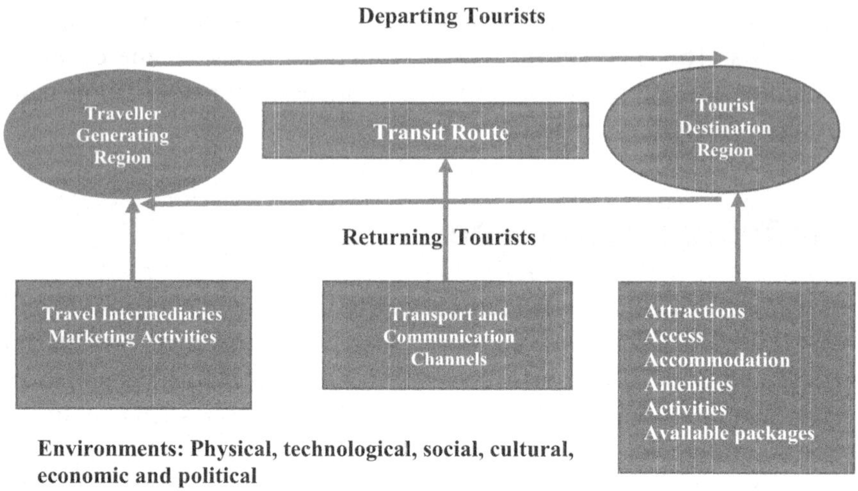

Fig. 1.1 The tourism system. (Source: Adapted from Leiper (1995))

management of the destination and the provision of quality service, which is vital for improving the images of destinations and motivating tourists to visit. The tourist destination functions as a "pull" factor and provides the space and resources for tourism activities at the destination. It provides tourists with a variety of attractions, activities and memorable travel experiences.

Elliott (1987) argued that the destination is multifaceted, including service, size, location, ownership, market and, most importantly, the actor-network relationship. Offering this multifaceted set of products and services is, therefore, a prerequisite for creating attractive destinations (Cracolici & Nijkamp, 2008). Ritchie and Zins (1978) identified eight dimensions of destination attractiveness, namely nature and beauty; culture and social characteristics; sport, recreation and education facilities; shopping and commercial facilities; infrastructure; the cost of living; attitudes towards tourists; and accessibility of the destination. Other factors that have been found to contribute towards the attractiveness of a destination include attractions and activities (Ritchie & Crouch, 2003), accessibility of a destination (Kim, 1998), the host community (Kim et al., 2012) and entertainment (Pullman & Gross, 2004).

However, for most emerging economies, their natural resource endowments have served as the main sources of attraction to tourists. Tang and Rochananond (1990) examined the attractiveness of Thailand as a tourist destination and found that natural beauty and climate, culture and social characteristics, cost of living and attitudes towards tourists were the more favourable attributes. Dickman (1997) coined the abbreviation 5As to describe the *destination mix*, namely attractions, access, accommodation, amenities and activities. However, a sixth element—available packages—has been added to the mix to make it 6As (Buhalis, 2000). *Attractions* are the locations, people, activities and things that draw visitors to a destination. Attractions could be in the form of natural and cultural landmarks, historical sites, monuments, zoos and game reserves, aquaria, museums and art galleries, gardens, architectural buildings, theme parks, sports facilities, festivals and activities, animals and people (Benckendorff, 2014). *Access* is the transport infrastructure and technology that enable people to reach their destination. *Accommodation* includes all forms of lodging or place of stay that caters for the needs of tourists regularly and at any time (Silaban et al., 2019). These include hostels, hotels, inns, boarding houses, rest houses and lodging houses. *Amenities* are the support facilities or infrastructure that can fulfil the needs and desires of tourists (Dewi et al., 2017). Singh et al. (2015) noted that there are four types of amenities, namely tourist infrastructure (accommodation facilities, facilities for arrivals servicing, tourist information and trails), paratourist infrastructure (transportation facilities—roads and transport points), local facilities (communal and public transport, trade and service facilities) and gastronomy and accompanying facilities (sport, leisure, entertainment). *Activities* can be described as tourism products that can build a portfolio or image for the country (Sharafuddin, 2015). *Available packages* refer to the availability of services bundled by intermediaries that promote the tourist destination. The United Nations World Tourism Organization (UNWTO, 2007) has indicated that the tourist destination is shaped by attractions, public and private amenities, accessibility, human resource, image and character, and price. The tourist destination concept has evolved tremendously since the 1990s with concepts like destination management, destination competitiveness, destination image and destination marketing.

Destination Marketing

There is now a recognition that what tourist destinations offer is not just a combination of various goods and services in a geographical unit, but rather the experience of visiting a specific destination (Fyall et al., 2019). Destination marketing is a type of marketing that seeks to promote a destination (town, city, region, country) with the view to increase the number of visitors (Lomanenko, 2018). Destination marketing can be defined as a proactive, visitor-centred approach to the economic and cultural development of a destination that balances and integrates the interests of visitors, service providers and the community (Wang, 2011). It is also defined as a management process through which the national tourist organisations and/or tourism enterprises identify their selected actual and potential tourists; communicate with them to ascertain and influence their wishes, needs, motivations, likes and dislikes, at the local, regional, national and international levels; and develop and adapt their tourist products with the view to achieving optimal tourist satisfaction, thereby fulfilling their objectives (3G Learning, 2015). It is about engaging with key players to drive awareness of the destination, thereby driving interest, increasing tourist arrivals at the destination. Destination marketing is a managerial process, demand-driven research, advertising and communication activity wherein the focus is on enticing potential external consumers.

Destination marketing is now acknowledged as a pillar of the future growth and sustainability of tourism destinations in an increasingly globalised and competitive market for tourists (UNWTO, 2011). Destination marketing as a management tool is a form of market-oriented strategic planning and hence as a strategic approach to place development rather than a promotional tool (Blumberg, 2007). Destination marketing covers all the activities from conception of ideas to realisation of profits. The scope of destination marketing can relate to the functions of marketing as well as the roles and responsibilities of the destination marketing organisations (DMOs) that revolve around customer satisfaction (Roshani, 2020) and as the main vehicle to compete and attract visitors to their distinctive place or visitor space (Pike & Page, 2014).

Destination Marketing Mix

Marketing activities play an essential role in determining the competitive position of a tourist destination; thus, various marketing mix elements need to be optimised. Marketing mix is a strategy of an organisation to do the tasks in offering its product and influence the demand (Bakhshi, 2007; Kumar & Sharma, 2009). Some experts (see, e.g. Kotler et al., 2008; Lilien, 1994) have recommended the use of marketing mix models for decision-making. This is also true for emerging tourist destinations. The classical marketing mix model consists of 4Ps, namely product, price, place and promotion (McCarthy, 1975). However, this has been criticised for not including uncontrollable external environmental factors. Thus, the marketing mix has been extended into 7Ps, adding people, process and physical evidence to the mix.

However, tourism destinations' distinctive characteristics have compelled some experts to propose other elements of the marketing mix, such as Morrison's (2009) 3Ps (partnership, packaging and programming). Similarly, Bao (2018) argues that partners, passion and the presentation should be added to the list to complete the traditional 7Ps. Nevertheless, tourism destination marketers need to find the right combination of marketing mix elements to offer suitable products to meet the right target markets and attract tourists. A case in point is the marketing mix strategy of Indonesia, which won the best destination marketing award in 2016 from Travel Weekly Asia. The marketing mix strategy of Indonesia involved the application of a promotion strategy by the Ministry of Tourism, which resulted in Indonesia winning. The strategy applied by the institution to promote Indonesia as an attractive tourism destination was Branding-Advertising-Selling (BAS), which was intensively conducted to significantly project the "Wonderful Indonesia" image between 2015 and 2016. This strategy increased Indonesia's position on the travel and tourism competitiveness index from not available (NA) to 47 out of 144 countries during the period. Indonesia's marketing strategy also focused on Bali, Jakarta and Batam-Bintan, applying the 3As, namely attractions, accessibility and amenity. The success of this strategy was followed by the design of an intensive promotional strategy to market ten new priority destinations to target 20 million tourists in 2019. Furthermore, as noted by Amran (2015), the government issued deregulation policies to attract international tourists, for instance, by (a) issuing a free visa regulation for international tourists from 169 countries, (b) permitting cruise ships to embark and disembark from Indonesian harbours and (c) issuing clearance approval for a yacht to enter Indonesian Territorial Waters (Clearance Approval for Indonesian Territory—CAIT).

Destination Marketing Organisations

Destination marketing organisations (DMOs) have become vital for coordinating the activities of emerging tourist destinations. The UNWTO (2007) categorised them into three, namely national tourism authorities (NTAs), regional provincial or state DMOs (RTOs) and local DMOs. Tourist destinations are generally managed by DMOs. According to Destinations International, a DMO is an organisation representing a particular destination and supporting local community's long-term development through travel and tourism strategies. Moreover, the UNWTO (2007) states that DMOs are responsible for managing and marketing destinations. The main functions of DMOs are marketing and ensuring that tourist destinations remain competitive (Pike & Ryan, 2004; Volgger & Pechlaner, 2014; Wang, 2008) by providing a collective platform for marketing destinations to tourists.

According to Timareva et al. (2015), DMOs play a role in organising and integrating all destination mix elements such as events, facilities, transportation, infrastructure and hospitality resources, based on long-term travel and tourism planning strategies. The strategy employed by a DMO usually includes image building, branding, marketing and communication, with the aim of attracting tourists to a destination. In the rural tourism destination context, Adeyinka-Ojo (2014) suggests that a DMO is a network management organisation that plays a role as network manager (Volgger & Pechlaner, 2014), especially in coordinating with industry and collaborating with tourism destination stakeholders (Wang, 2008). The primary goals of a DMO are to provide mutual benefits for

Table 1.1 Summary of the activities of the DMO categorised as either EDM or IDD

External destination marketing activities	Internal destination development activities
Web marketing	Visitor management
Events, conferences and festivals	Information/research
Cooperative programmes	Coordinating tourism stakeholders
Direct mail	Crisis management
Direct sales	Human resources development
Sales blitzes	Finance and venture capital
Trade shows	Resource stewardship
Advertising	Quality of the visitor experience
Familiarisation tours	
Publications and brochures	

Source: Arbogast et al. (2017)

various stakeholders, market the tourism products of a destination, position the destination, improve the host community's well-being, build brands and ensure tourists' loyalty towards tourist destinations. To achieve these goals, DMOs need to collaborate and share the same visions with the host community, an important stakeholder that creates a tourism destination experience (Bhat & Gaur, 2012). Overall, DMOs need to ensure that internal communication and socialisation run effectively; thus, the stakeholders hold on to the shared destination vision and are committed to developing tourist destinations (Bregoli, 2013).

The responsibilities of DMOs should include both an "external destination marketing" (EDM) function and an "internal destination development" (IDD) function. The EDM function is meant to include all activities aimed at attracting visitors to the destination. These activities have an external orientation in that they aim to influence the actions of persons outside the destination. IDD activities are described as encompassing all other forms of activities (apart from marketing) undertaken by DMOs to develop and maintain tourism in the destination. The function of destination development is internal in that the activities are aimed at initiatives internal to the destination (Arbogast et al., 2017) (Table 1.1).

DMOs as a vehicle for tourism destination management and development are unlikely to be able to claim too much responsibility for destination management, but they can play an important part in the management of the destination product (Blumberg, 2007). DMOs seek to attract more tourists to a destination. Today, DMOs, according to Čorak and Živoder (2017), have a complex task to ensure sustainable destination development through optimal use of resources and welfare of the local community and achieve competitiveness in a fast-changing social and economic climate. Furthermore, the DMO, as the leading entity in the tourism destination, has a crucial role in promoting a greater engagement of the tourism sector, its industries, as well as policy and decision-makers with sustainable development. DMOs should align their policies and actions with the 2030 Agenda for Sustainable Development and the Sustainable Development Goals (SDGs) (UNWTO, 2019). This will allow them to maximise tourism's contribution to all the 17 Goals and in particular to the SDGs where tourism is specially included: (a) Goal 8: decent work and economic growth; (b) Goal 12: sustainable production and consumption; and (c) Goal 14: life below water. Within this context, DMOs are called upon to incorporate the 17 SDGs in their plans and policies and to monitor their implementation on an ongoing basis in order to ensure compliance with the final goal of maximising the contribution of tourism destination to the SDGs.

The responsibility of a DMO is not only limited to planning, marketing and promotion but also coordinating various developmental activities (Presenza et al., 2005; Ritchie & Crouch, 2003). For destinations to remain competitive, DMOs must play an important role of marketing, promoting, building favourable images and providing information about tourist destinations (Prideaux & Cooper, 2003). Pike (2004) stated that the function of a DMO relates to human resources, politics, destination policy and strategy, finance and budgeting, destination branding, tourism assets and attractions, destination communication, promotion and monitoring of performance. Heath and Wall (1992) also have a different perspective on the function of DMOs, stating that the main tasks of DMOs are strategy formulation, promotion of stakeholders' interest, product development and marketing.

DMOs' success indicators are divided into two aspects: external and internal performances. Marketing is part of the external performance, while internal performance could be measured by coordination of destination stakeholders (Ritchie & Crouch, 2003; Presenza et al., 2005) to create quality, satisfying and memorable tourist experiences (Gartrell, 1988). The success can be achieved by managing networking, transparency, resources and professionalism, which are essential for effective coordination and integration of the DMO and other stakeholders (Volgger & Pechlaner, 2014). In the fourth industrial revolution era, technology development encourages DMOs to use online platforms such as websites and social media to build the image of and promote destinations (Molinillo et al., 2018). In Singapore, for instance, the DMO and tourism board aim to change the image of the country, which is one of the biggest hubs in Southeast Asia, from a "stopover" destination to a real tourist destination. To achieve this goal, a strategy that involves sharing real stories of residents and local experts about the attractions in Singapore has been employed. The campaign is conducted on microsites, exposing potential tourists to cultural attractions including food, and the uniqueness they will experience when they visit Singapore. In this case, the DMO and the tourism board, with the help of media experts, created an original and memorable experience for tourists visiting Singapore.

Nevertheless, some DMOs, especially in emerging economies, are confronted with a number of challenges. Perhaps, the greatest challenge is the fragmented nature of the tourism industry, with varied interest groups, which makes coordination of the activities of various stakeholders towards the realisation of shared goals a very daunting task (Laws et al., 2011). There are also problems with weak institutional structures and government bureaucracy, which result in inefficiencies and delays in decision-making. Waitt (1996), for instance, examined the destination marketing of Korea post-Olympic Games 1988 and found that the government bureaucracy was an obstacle to Korea becoming the leading destination, even though the country had received attention worldwide from the successful hosting of the Olympic Games.

Features of Emerging Economies

Though there is no commonly accepted definition of emerging markets (Arnold & Quelch, 1998), they could be described as low- or middle-income economies with high economic growth potential. They are countries that are transitioning from the developing phase to the developed phase. Emerging economies are also known as developing countries. Emerging economies have shifted from agriculture and export of raw materials to industrialisation and have invested in a more productive capacity. Several countries fall into this category. However, in the context of this book, they include the newly industrialised economies in East Asia such as Korea, Taiwan, Hong Kong and Singapore (Asian NIEs); Brazil, Russia, India and China (BRICS); Mexico, Indonesia, Nigeria and Turkey (MINT); Colombia, Indonesia, Vietnam, Egypt, Turkey and South Africa (CIVETS); Countries of Central and Eastern Europe, including the former Soviet Union, China and Mongolia (Transition economies); and the new emerging economies of Africa, namely Botswana, Ghana, Kenya, Mozambique, Nigeria, Tanzania, Uganda and Zambia. Emerging economies differ significantly from developed countries in terms of their cost and factor conditions; their stages of economic, social and political development; as well as their business systems (Enderwick, 2007).

Sunje and Civi (2008) describe an emerging market as a market within the political borders of an independent state that has a lower level of economic development, transitional economy and society, a high rate of economic growth and a huge future growth potential.

Atadil et al. (2017) describe emerging markets in the context of tourism as both emerging sources and destination markets. Emerging economies exhibit a lot of potentials, including a rapid pace of economic development, high growth potential, trade liberalisation and a free market economy. These economies have become increasingly important in the global economy in recent years. Emerging market economies accounted for close to 70% of global output growth and just over 70% of global consumption growth during 2010–2015 (International Monetary Fund [IMF], 2014). It has been projected that by 2035, the GDP of emerging markets will permanently exceed that of all advanced markets (Wilson & Purushothaman, 2003). It has been forecasted that by 2050, the BRIC economies, namely Brazil, Russia, India and China, will overtake developed countries such as the United States, Japan, Germany, the UK, France and Italy (Wilson & Purushorthaman, 2003). This has been attributed to the strong competitive challenge from domestic firms in these emerging markets (Ramamurti & Singh, 2009).

In spite of this, emerging economies are confronted with many challenges which hinder them from achieving their full economic and tourism potentials. These include informality and less developed government and regulatory infrastructure (Marquis & Qian, 2014; Marquis et al., 2011); lapses in institutional frameworks (Meyer & Peng, 2016); inadequate commercial and physical infrastructure (Sunje & Civi, 2008); susceptibility to external conflicts, coups and internal tensions (Hiatt & Sine, 2014); less transparency; state intervention and high volatility of key economic, political and institutional variables (Xu & Meyer, 2013); as well as a youthful population, expanding workforce and rapid urbanisation (Marquis & Raynard, 2015). These socio-economic challenges, coupled with the impacts of crises, have significantly impacted the efforts of DMOs in their efforts at competing in the global tourism industry. Technological advancements, structural changes in the tourism market, economic downturn, war and terrorist threats, climate change, natural disasters as well as epidemics and pandemics have resulted in crises that affect DMOs in fundamental ways (Gretzel et al., 2006).

Tourism Prospects of Emerging Economies

Tourism has become one of the most important sectors in many emerging markets (Singh, 1997). For these economies, tourism has become a major source of economic growth and foreign income (Norwak et al., 2003). The UNWTO (2015) projects that arrivals in emerging destinations between 2010 and 2030 will increase at twice the rate of the advanced economies, and this will account for a 57% share of the market. Africa, the Middle East, and especially Asia and the Pacific regions are expected to grow faster. Compared to the emerging economies, the growth rate in tourist arrivals in mature destinations in the developed economies of Northern and Western Europe, and North America will be comparatively slower during the coming two decades.

As a result of these prospects, many governments of emerging economies are leaving no stone unturned at developing their tourism industries. This is also because emerging markets enjoy a huge cost and price advantage from tourism (Enderwick & Nagar, 2011). Tourism growth globally, over recent years, has largely been driven by emerging economies (Kester & Croce, 2011).

Tourism in emerging markets has drawn increasing attention (Cohen et al., 2014). Emerging tourist destinations have increasingly become attractive to tourists from the key source markets in North America and Europe due to their natural and cultural attractions, which offer suitable alternatives to the traditional attractions in advanced countries. Many destinations in South East Asia, Africa and Latin America offer cultural, heritage and natural attractions, which appeal to Western tourists who are in search of novelty and exotic experiences. Hitchcock et al. (2009) describe this as the search for something different, with tourists prioritising new natural and cultural environments. Emerging economies have a comparative advantage in their natural endowments, including climate, wildlife and cultural resources. They offer a rich supply of attractions and natural resources that appeal to the main source markets in Europe and America. According to Laws (1996), the shift in tourists' interests towards destinations in Southeast Asia, Africa and the Pacific is a reflection of changes in tourists' preferences. Emerging economies must capitalise on these shifts in preferences by leveraging their competitive advantage in cultural and natural attractions.

Challenges of Marketing Emerging Destinations

Despite their huge potentials in the global tourism market, emerging economies are confronted with a myriad of challenges that impede successful destination marketing. Tourism-related challenges confronting emerging economies include the absence of a national tourism organisation (NTO), poor coordination among institutions, non-prioritisation of the tourism industry by government, insufficient use of ICT, lack of proper marketing and promotion, inadequate human resources and poor tourism infrastructure (Arshad et al., 2018; Kimbu, 2011). According to Mitchell (2006), the most apparent challenges facing the tourism industry in developing economies include poor infrastructure; poor product development and management; poor marketing; poor linkages within the local economy; poor institutional and technical capabilities; and shortage of suitable, qualified and skilled personnel. The preponderance of crises, including financial crisis, natural disasters and epidemics in emerging markets, is a major drawback to destination marketing. These crises do not only render destinations unsafe for travel but also cast a slur on the image of such destinations. Over the past decade, emerging economies have had a disproportionate share of crises, including the tsunami in the Indian Ocean basin, SARS pandemic in the Asia Pacific region, Ebola epidemic in West Africa and political unrests in North Africa and the Middle East.

The technological dependency of emerging markets (Wu & Pangarkar, 2006) is also a major constraint to destination marketing, especially in the COVID-19 era when almost everything has been moved to the virtual space, and there has been the emergence of virtual tourism and smart destinations. The inadequacy of ICT infrastructure in most emerging economies renders them uncompetitive in the global tourism market.

There is also a problem with the images and the tourism product offerings of some destinations in emerging economies. Third World destinations face major difficulties in creating favourable destination images and are often portrayed negatively in international tourism marketing, especially in terms of context and content (Dieke, 2010; Echtner & Prasad, 2003). Images influence actual travel behaviour (O'Leary & Deegan, 2005), so the association of destinations in the developing world with crude and negative images tend to affect tourist arrivals in those destinations. Finally, emerging economies have not made significant strides in differentiating their tourism product offerings (Shani et al., 2009). Daye (2010) laments that destinations such as the Dutch Caribbean islands are mainly undifferentiated in terms of their tourism product offerings and also in terms of general consumer awareness of those destinations. However, destination positioning and clarity of destination image are factors that significantly influence the competitiveness of tourist destinations (Dwyer & Kim, 2003).

Structure of the Book

The book is composed of 13 chapters, further organised under 3 sections (Parts I, II and III). Part A, which is entitled "Destination Attributes and Features", is composed of the first four chapters. In Chap. 1, Ishmael Mensah, Mohd Raziff Jamaluddin, Vanessa Gaffar, Gina Alcoriza, Kandappan Balasubramanian and S. Mostafa Rasoolimanesh introduce readers to the concepts, prospects and challenges of destination marketing in emerging economies. Vanessa Gaffar examines the marketing mix strategies of emerging tourist destinations using Indonesia as a case study in Chap. 2. This is followed by an analysis of special events as destination marketing strategies in emerging economies in Asia by Raziff Jamaluddin in Chap. 3. The section ends with an assessment of the competitiveness of emerging tourist destinations by Emmanuel Gamor and Ishmael Mensah in Chap. 4.

Part B, which comprises four chapters, is labelled "Travel Behaviour and ICT Applications in Destination Marketing". The section begins with an analysis of the travel decision-making process and motivations for travel to emerging tourist destinations by Benjamin Appiah Osei in Chap. 5. In Chap. 6, Eunice Amissah, Evelyn Addison-Akotoye and Sarah Blankson-Stiles-Ocran examine the issue of service quality, tourist satisfaction and destination loyalty in emerging economies. Christopher Mensah, Richmond Sakyi and Mavis Adjoa Forson provide an overview of destination distribution systems and disintermediation in emerging economies in Chap. 7. The section ends with an examination of the progress of information technology's applications in destination marketing in emerging economies by Kandappan Balasubramanian and S. Mostafa Rasoolimanesh in Chap. 8.

Part C is entitled "Destination Marketing Process and Activities". There are five chapters in this section (Chaps. 9, 10, 11, 12 and 13). The section begins with an examination of the approaches to market segmentation, targeting and positioning of emerging tourist destinations, with reference to the Maribojoc Bohol area of the Philippines by Maria Criselda G. Badilla in Chap. 9. In Chap. 10, Ishmael Mensah examines the use of homecoming events for diaspora tourism promotion in emerging economies, using the Year of Return 2019 campaign in Ghana. The impacts of destination branding slogans on tourist arrivals in the Philippines is analysed by Reil G. Cruz and Joy Sheelah B. Era in Chap. 11. This is followed by an evaluation of the opportunities and challenges of the "One Belt One Road" as a globalisation strategy for ASEAN destinations in Chap. 12. The final chapter (Chap. 13) completes the section with an examination of the issue of marketing emerging tourist destinations during crises and pandemics by Ishmael Mensah.

Conclusion

Tourism destinations are becoming competitive as more and more destinations look to tourism as the new economic generator, replacing activities in agriculture, mining and manufacturing (Goeldner & Ritchie, 2006). This is especially the case with emerging economies that have to compete with the well-established destinations in North America, Australia and Europe. The fact is that tourist destinations in emerging economies are confronted with a number of issues, including crises like terrorism, political unrest, natural disasters and epidemics, as well as institutional and structural deficiencies, which tend to affect the image, attractiveness and marketing efforts of DMOs of such destinations. Addressing these challenges calls for a well-coordinated management and marketing strategy. The tourist destination is very dynamic and new issues have emerged in contemporary times, which have implications for destination marketing in emerging economies. These include the emergence of new tourist destinations, destination branding, disintermediation in the distribution system, application of ICT, changes in consumer behaviour and collaboration. This book addresses these issues by explaining the concepts and drawing practical examples from emerging economies. This book aims to highlight the destination marketing challenges, best practices and strategies relevant to emerging economies.

References

3G Elearning. (2015). *Destination marketing and management.* 1G Elearning FZ LLC, UAE.

Adeyinka-Ojo, S. F., Khoo-Lattimore, C., & Nair, V. (2014). A framework for rural tourism destination management and marketing organisations. *Procedia—Social and Behavioral Sciences, 144*, 151–163.

Amran, F. (2015). *How would you promote Indonesia as tourist destination?* https://addiction.id/2015/03/30/how-would-you-promote-indonesia-as-a-tourist-destination/

Arbogast, D., Deng, J., & Maumbe, K. (2017). DMOs and rural tourism: A stakeholder analysis the case of Tucker County, West Virginia. *Sustainability, 9*(10), 1813.

Arnold, D. J., & Quelch, J. A. (1998). New strategies in emerging markets. *MIT Sloan Management Review, 40*(1), 7.

Arshad, M. I., Iqbal, M. A., & Shahbaz, M. (2018). Pakistan tourism industry and challenges: A review. *Asia Pacific Journal of Tourism Research, 23*(2), 121–132.

Atadil, H. A., Sirakaya-Turk, E., & Altintas, V. (2017). An analysis of destination image for emerging markets of Turkey. *Journal of Vacation Marketing, 23*(1), 37–54.

Bakhshi, B. (2007). *The impact of tourism marketing mix on attracting domestic and foreign tourists in Kish Island.* Master Thesis, Tehran, Islamic Azad University.

Bao, H. (2018). Marketing of tourism services/experiences. In M. Sotiriadis (Ed.), *The Emerald handbook of entrepreneurship in tourism, travel and hospitality* (pp. 261–275). Emerald.

Beirman, D. (2020). *Restoring tourism destinations in crisis: A strategic marketing approach.* Routledge.

Benckendorff, P. (2014). Attraction, tourism. In J. Jafari & H. Xiao (Eds.), *Encyclopaedia of Tourism.* Springer.

Bhat, S., & Gaur, S. S. (2012). Managing diverse stakeholders in the context of destination marketing. *Worldwide Hospitality and Tourism Themes, 4*(2), 185–202.

Blumberg, K. (2007). Tourism destination marketing—A tool for destination management? A case study from Nelson/Tasman Region, New Zealand. *Asia Pacific Journal of Tourism Research, 10*(1), 45–57.

Bregoli, I. (2013). Effects of DMO coordination on destination brand identity: A mixed-method study on the city of Edinburgh. *Journal of Travel Research, 52*(2), 212–224.

Buhalis, D. (2000). Marketing the competitive destination of the future. *Tourism Management, 21*, 97–116.

Cohen, S. A., Prayag, G., & Moital, M. (2014). Consumer behaviour in tourism: Concepts, influences and opportunities. *Current Issues in Tourism, 17*(10), 872–909.

Čorak, S., & Živoder, S. B. (2017). Tourism destination and DMO transformation. In *Evolution of destination planning and strategy* (pp. 99–118). Palgrave Macmillan.

Cracolici, M. F., & Nijkamp, P. (2008). The attractiveness and competitiveness of tourist destinations: A study of southern Italian regions. *Tourism Management, 30*, 336–344.

Daye, M. (2010). Challenges and prospects of differentiating destination brands: The case of the Dutch Caribbean islands. *Journal of Travel & Tourism Marketing, 27*(1), 1–13.

Dewi, N. I. K., Astawa, I. P., Siwantara, I. W., & Bagus, I. G. A. (2017). Strategy development: Case study on two cultural villages in Bali. *Journal of Advances in Humanities and Social Sciences, 3*(6), 293–302.

Dickman, S. (1997). *Tourism: An introductory text*. Hodder Education.

Dieke, P. (2010). Tourism and political change in South Africa. In R. Butler & W. Suntikul (Eds.), *Tourism and political change* (pp. 120–132). Goodfellows.

Dwyer, L. P., & Kim, C. (2003). Destination competitiveness: Determinants and indicators. *Current Issues in Tourism, 6*(5), 369–414.

Echtner, C. M., & Prasad, P. (2003). The context of third world tourism marketing. *Annals of Tourism Research, 30*(3), 660–682.

Elliott, J. (1987). Government management of tourism—a Thai case study. *Tourism Management, 8*(3), 223–232.

Enderwick, P. (2007). *Understanding emerging markets: China and India*. Routledge.

Enderwick, P., & Nagar, S. (2011). The competitive challenge of emerging markets: The case of medical tourism. *International Journal of Emerging Markets, 6*(4), 329–350.

Fyall, A., Legoherel, P., Frochot, I., & Wang, Y. (2019). *Marketing for tourism and hospitality collaboration, technology and experiences*. Routledge.

Gartrell, R. B. (1988). *Destination marketing for convention and visitor Bureaus*. Kendall.

Goeldner, C. R., & Ritchie, R. B. (2006). *Tourism*. John Wile & Sons.

Gretzel, U., Fesenmaier, D. R., Formica, S., & O'Leary, J. T. (2006). Searching for the future: Challenges faced by destination marketing organizations. *Journal of Travel Research, 45*(2), 116–126.

Heath, E., & Wall, G. (1992). *Marketing tourism destinations: A strategic planning approach*. John Wiley.

Hiatt, S. R., & Sine, W. D. (2014). Clear and present danger: Planning and new venture survival amid political and civil violence. *Strategic Management Journal, 35*(5), 773–785.

Hitchcock, M., King, V. T., & Parnwell, M. (Eds.). (2009). *Tourism in southeast Asia: Challenges and new directions*. Nias Press.

International Monetary Fund. (2014). *Emerging markets in transition: Growth prospects and challenges* (Staff Discussion Note 14/06). International Monetary Fund.

Kester, J., & Croce, V. (2011). Tourism development in advanced and emerging economies: What does the travel and tourism competitiveness index tell us. In J. Blanke & T. Chiesa (Eds.), *The Travel and tourism competitiveness report (2011)* (pp. 45–52). World Economic Forum.

Kim, H. B. (1998). Perceived attractiveness of Korean destinations. *Annals of Tourism Research, 25*(2), 340–361.

Kim, J., Ritchie, J. R. B., & McCormick, B. (2012). Development of a scale to measure memorable tourism experiences. *Journal of Travel Research, 51*(1), 12–25.

Kimbu, A. N. (2011). The challenges of marketing tourism destinations in the central African subregion: The Cameroon example. *International Journal of Tourism Research, 13*(4), 324–336.

Kotler, P., Armstrong, G., Wong, V., & Saunders, J. (2008). *Principles of marketing* (12th ed.). Pearson Educational International.

Kumar, D. A., & Sharma, H. (2009). Services marketing mix in library & information centers. *ICAL, Advocacy & Marketing, 2*, 456–460.

Laws, D. (1996). *The context of international tourism*. Sheffield Hallam University.

Laws, E., Richins, H., Agrusa, J., & Scott, N. (2011). *Tourist destination governance: Practice, theory and issues*. CAB International.

Leiper, N. (1995). *Tourism management*. RMIT Press.

Lilien, G. L. (1994). Marketing models: Past, present and future. In *Research traditions in marketing* (pp. 1–26). Springer.

Lomanenko, S. (2018, July 12). What is destination marketing. https://www.promodo.com/blog/what-is-destination-marketing/

Marquis, C., & Qian, C. (2014). Corporate social responsibility reporting in china: Symbol or substance? *Organization Science, 25*(1), 127–148.

Marquis, C., & Raynard, M. (2015). Institutional strategies in emerging markets. *Academy of Management Annals, 9*(1), 291–335.

Marquis, C., Zhang, J., & Zhou, Y. (2011). Regulatory uncertainty and corporate responses to environmental protection in China. *California Management Review, 54*(1), 39–63.

McCarthy, J. (1975). *Basic marketing: A managerial approach*. Richard D. Irwin.

Meyer, K. E., & Peng, M. W. (2016). Theoretical Foundations of Emerging Economy Research. *Journal of International Business Studies, 47*(1): 3–22.

Mitchell, J. (2006). *Can tourism reduce poverty in Africa?* (ODI Briefing Paper). http://dlc.dlib.indiana.edu/dlc/bitstream/handle/10535/4569/tourism%20business.pdf?sequence=1

Molinillo, S., Liébana-Cabanillas, F., Anaya-Sánchez, R., & Buhalis, D. (2018). DMO online platforms: Image and intention to visit. *Tourism Management, 65*, 116–130.

Morrison, A. M. (2009). *Hospitality and travel marketing* (4th ed.). Cengage Learning.

Norwak, J. J., Sahli, M., & Sgro, P. M. (2003). Tourism, trade, and welfare. *Pacific Economic Review, 8*(3), 245–258.

O'Leary, S., & Deegan, J. (2005). Career progression of Irish tourism and hospitality Management graduates. *International Journal of Contemporary Hospitality Management, 17*, 421–432.

Pike, S. (2004). *Destination marketing organisations*. Elsevier.

Pike, S., & Page, S. (2014). Destination marketing organizations and destination marketing: A narrative analysis of the literature. *Tourism Management, 41*, 1–26.

Pike, S., & Ryan, C. (2004). Destination positioning analysis through a comparison of cognitive, affective and conative perceptions. *Journal of Travel Research, 42*(4), 333–342.

Presenza, A., Sheehan, L., & Ritchie, J. R. B. (2005). Towards a model of the roles and activities of destination management organizations. *Journal of Hospitality, Tourism and Leisure Science, 3*(1), 1–16.

Prideaux, B., & Cooper, C. (2003). Marketing and destination growth: A symbiotic relationship or simple coincidence? *Journal of Vacation Marketing, 9*(1), 35–51.

Pullman, M. E., & Gross, M. A. (2004). Ability of experience design elements to elicit emotions and loyalty behaviors. *Decision Sciences, 35*(3), 551–578.

Ramamurti, R., & Singh, J. V. (Eds.). (2009). *Emerging multinationals in emerging markets.* Cambridge University Press.

Ritchie, J. B., & Zins, M. (1978). Culture as determinant of the attractiveness of a tourism region. *Annals of Tourism Research, 5*(2), 252–267.

Ritchie, J. R. B., & Crouch, G. I. (2003). *The competitive destination: A sustainable tourism perspective.* CABI.

Roshani, R. (2020). *Scope of marketing, seven major scopes of marketing.* https://www.businessmanagementideas.com/marketing/scope-of-marketing-7-major-scope-of-marketing/19727

Shani, A., Wang, Y., Hudson, S., & Gil, S. M. (2009). Impacts of a historical film on the destination image of South America. *Journal of Vacation Marketing, 15*(3), 229–242.

Sharafuddin, M. A. (2015). Types of tourism in Thailand. *e-Review of. Tourism Research, 12*(3/4), 210–219.

Silaban, P. H., Pasaribu, A., & Silalaho, A. D. K. (2019). The influence of human aspect of accommodation and destination on tourist satisfaction. *International Journal of Innovative Technology and Exploring Engineering, 8*(7), 985–995.

Singh, A. (1997). Asia Pacific tourism industry: Current trends and future outlook. *Asia Pacific Journal of Tourism Research, 2*(1), 89–99.

Singh, A., Saini, G. K., & Majumdar, S. (2015). Application of social marketing in social entrepreneurship: Evidence from India. *Social Marketing Quarterly, 21*(3), 152–172.

Sunje, A., & Civi, E. (2008). Emerging markets: A review of conceptual frameworks. In *Proceedings of 1st international joint symposium on business administrator: Challenges for business administrators in the New Millennium* (pp. 203–216). Silesian Unviersity.

Tang, J., & Rochananond, N. (1990). Attractiveness as a tourist destination: A comparative study of Thailand and selected countries. *Socio-economic Planning, 24*(3), 34–43.

Timareva, S., Arabska, E., & Shopova, I. (2015). Role of destination management and marketing organizations in regional development. *Trakia Journal of Sciences, 13*(Suppl 1), 96–102.

Tinsley, R., & Lynch, P. (2001). Small tourism business networks and destination development. *International Journal of Hospitality Management, 20*(4), 367–378.

United Nations World Tourism Organisation. (2007). *A practical guide to destination management.* UNWTO.

UNWTO. (2011). *United Nations World Tourism Organization (UNWTO) policy and destination management.* https://www.unwto.org/policy-destination-management

UNWTO. (2015). *UNWTO tourism highlights, 2015 Edition*. UNWTO.

UNWTO. (2019). *UNWTO guidelines for institutional strengthening of destination management organizations (DMOs).* UNWTO.

Volgger, M., & Pechlaner, H. (2014). Requirements for destination management organizations in destination governance: Understanding DMO success. *Tourism Management, 41*, 64–75.

Waitt, G. (1996). Marketing Korea as an international tourist destination. *Tourism Management, 17*(2), 113–121.

Wang, Y. (2008). Collaborative destination marketing: Roles and strategies of convention and visitors bureaus. *Journal of Vacation Marketing, 14*(3), 191–209.

Wang, Y. C. (2011). *Destination marketing and management: Theories and applications.* CABI.

Wilson, D., & Roopa, P. (2003). *Dreaming with BRICs: The path to 2050* (Global Economics Paper No. 99). Goldman Sachs.

Wu, J., & Pangarkar, N. (2006). Rising to the global challenge: Strategies for firms in emerging markets. *Long Range Planning, 39*(3), 295–313.

Xu, D., & Meyer, K. E. (2013). Linking theory and context: Strategy research in emerging economies. *Journal of Management Studies, 50*(7), 1322–1346.

2

Marketing Mix Strategies of Emerging Tourist Destinations: The Case of Indonesia

Vanessa Gaffar

Introduction

The tourism market is a set of relationships between demands and offers, focused on the exchange of goods and services, utilising money (Ciriković, 2014). The differences between tourism marketing and other services are (a) products provided are transportation, hospitality and recreational experiences; (b) tourists should travel to the place of production (destination) instead of moving the product to the customer; and (c) the portion of the time and expenses spent on travel are mostly for recreation and tourism experience purposes (Middleton et al., 2009; Srinivasan, 2009). To provide optimum experiences to the tourism market, tourism marketers should employ various combinations of marketing mix elements tailored to the requirements of the tourism target market (Pomering et al., 2011). The success of the tourism destination depends on its ability to

V. Gaffar (✉)
Universitas Pendidikan Indonesia, Bandung, Indonesia
e-mail: vanessa@upi.edu

I. Mensah et al. (eds.), *Marketing Tourist Destinations in Emerging Economies*, Palgrave Studies of Marketing in Emerging Economies,
https://doi.org/10.1007/978-3-030-83711-2_2

manage and adapt the marketing mix to suit the demands of the tourism market. In the tourism industry, marketers can combine various marketing mix elements to differentiate their market offerings, which eventually form the desired brand positioning that leads to a competitive position (Ciriković, 2014).

The concept of the marketing mix was first introduced by Neil Borden in 1953. He proposed 12 elements of the marketing mix, namely product planning, pricing, branding, channels of distribution, personal selling, advertising, promotions, packaging, displays, servicing, physical handling, and fact-finding and analysis. However, only four elements known as the 4Ps were further developed by McCarthy (1960), namely product, price, place and promotion. Then, in 1981, Booms and Bitner added 3Ps to the marketing mix elements for service marketing, namely people, process and physical evidence. The adaptation of the service marketing mix resulted in a specific marketing mix for different service industries, including tourism, that has distinctive characteristics. Therefore, Morrison (2009) added 3Ps for tourism, namely partnership, packaging and programming. Apart from the existing elements of the marketing mix, it is essential for tourism destination marketers to find the appropriate marketing mix combination, which enables them to offer products that meet the requirements of their target markets and attract the attention of tourists. In this chapter, ten elements of the marketing mix, namely product, price, place, promotion, people, process, physical evidence, partnership, packaging and programming, are discussed in relation to Indonesia as a tourist destination.

Marketing Mix Elements for Emerging Tourist Destination

As mentioned earlier, the tourism industry has different characteristics from other service industries and, therefore, possesses specific marketing mix elements. This section explains each element of the tourism marketing mix and its application, particularly in the context of emerging tourist destinations. Emerging tourist destinations refer to tourism destinations in emerging economies, a term given to developing countries that have some characteristics of developed markets but do not fully meet their standards. Emerging economies have relatively rapid economic growth, supported by their tourism industry, which is known as an emerging tourism destination (Benjamin, 2010).

Product

A product is anything to be offered to a market that might satisfy a want or need (Kotler & Keller, 2006). In the tourism industry, destinations managed by the government or the private sector can be categorised as products. The destination offers a composite product that includes elements such as attractions and accommodation services, and tours can also be classified as a product (Pomering et al., 2011). Thus, the tourism product is an assembly of tourism attractions (Nouri & Soltani, 2015), complex physical resources, activities and services, which, when combined with natural resources, will create the desired experiences (Ciriković, 2014). According to Fyall and Garrod (2005), a tourism product is a combination of attractions and its environment, services and facilities, accessibility and imagery perception provided by the destination.

Tourism products should be tailored to meet the wants and needs of tourists. One of the key objectives of destination marketing organisations (DMOs) is destination positioning, which is the impressions of the destination that exist in the minds of customers and its uniqueness from competing destinations. In terms of international tourism, the product is a destination that is rich in natural, cultural or some other features that serve as an attraction. Therefore, national protected areas and national heritage are important for attracting international tourists. Destinations are tourism products that offer various experiences to consumers, comprising several components in the destination mix (Buhalis & Costa, 2005), namely attractions (natural, man-made, artificial, purpose-built, heritage and special events), accessibility (entire transportation system comprising routes, terminals and vehicles), amenities (accommodation and catering facilities, retailing and other tourist services), available packages (prearranged packages by intermediaries and principals), activities (all activities available at a destination and those to be carried out during the visit) and ancillary services (various services used by tourists such as banks, telecommunications, post, newsagents, hospitals, etc.).

In Indonesia, for instance, sports tourism is a developing tourism product that involves various sports, followed not only by professional athletes but also by amateurs. Existing natural potentials such as mountains, cliffs, caves, waterfalls, tropical forests, rivers and lakes provide opportunities for various sports activities, ranging from mountain climbing, rafting, rock climbing, mountain running, paragliding, skydiving, bicycle racing, surfing and other sports. Besides, Indonesia has been able to hold various international sports competitions, such as Bintan Ironman, Tour de Singkarak, Sumbawa Pro Surfing and MotoCross/ MXGP. These international sports events help in the development of sports tourism that creates strategic collaboration between tourism and sports and has high economic value for the country, regions and local communities. Through sports tourism, tourists also directly experience the local culture and traditions. Several sports tourism destinations in Indonesia are popular, such as the Raja Ampat and Wakatobi areas as a paradise for international diving, Bintan Island for racing, cycling and swimming, Borobudur for the heritage marathon, Singkarak for the Bike Marathon Tour de Singkarak, Nias as surf spots that bring in international surfers and Bali as the venue for various yoga festivals.

Price

Price is the amount of money charged for obtaining, consuming and using a product (Fyall & Garrod, 2005) that affect consumer demand and choice among competing brands (Avlonitis & Indounas, 2007; Kotler & Keller, 2006). Pricing of services is more difficult to implement than the pricing of products (Amzad et al., 2012), as is the case in the tourism industry. Pricing tourism products is a very sensitive issue, considering the fact that, during a period of intense competition, a high price may decrease sales and affect total income, but the price should be able to cover production costs and generate profits (Ciriković, 2014). Factors such as the increasing level of competition, changes in demand and tastes of tourists, and increase in unemployed extensive capacities should be considered by companies engaged in the tourism sector in determining the appropriate price (Ciriković, 2014).

The price charged to tourists usually includes transportation service fees to and from a destination and other costs such as accommodation, tour services, food and beverage, and entertainment. The price charged for foreign tourists should include currency conversion following the exchange rate and the comparative price of a similar destination in their home country (Dwyer & Forsyth, 2011). Price is influenced by several factors, such as the state of the economy, as well as the monetary and fiscal policies of a country. It can be concluded that price is one of the factors of resource allocation and positioning of a tourism organisation (Ciriković, 2014). The cost of travel to a tourism destination is determined by several factors such as inflation and the overall price level, exchange rates, wage costs, taxes, infrastructure and environmental costs, productivity performance of the tourism industry, export booms and Dutch disease (Dwyer & Forsyth, 2011). According to Dwyer and Forsyth (2011), the price factor in the context of tourism is the tourism cost charged to visitors, including the level of differentiation of the price of goods and services across different countries, for example airline tickets, airport taxes, tour services, food and entertainment, fuel prices, taxes in the country (which can be passed on to tourists) and the relative cost of accommodation. Price competitiveness is one of the key indicators for assessing the competitiveness of tourist destinations. In this regard, the Association of Southeast Asian Nations (ASEAN) as a whole has a price advantage compared with parts of the rest of Asia and elsewhere in the world (World Economic Forum, 2012). However, within the ASEAN, Singapore was adjudged the most competitive and surpassed only by Brunei and Indonesia (Henderson, 2015).

The current pandemic situation has brought travel and tourism to a standstill. Thus, the World Travel and Tourism Council has stated that the losses from the tourism industry in 2020 were around $2.7 trillion. As an attempt to revive the tourism industry, destinations are focusing on domestic tourists. This requires several strategies, including a price adjustment strategy. In Indonesia, the Ministry of Tourism and Creative Economy is recently collaborating with Garuda Indonesia Airline and the hotel industry to prepare packages offered at a low price to attract tourists, particularly domestic tourists, when the destination is reopened. The concept of staycation has become popular during the time of the COVID-19 pandemic since people have been reluctant to travel far. Thus, they have been directed to travel to places not far from their homes, such as spending their time staying and having recreation in hotels not far from their homes. Offering various packages at low prices for overnight stays becomes the main attraction for local tourists.

Place

The next element of the marketing mix is place. Kotler et al. (2017) stated that a place in the concept of tourism marketing is a place that tourists can visit for simple pleasures that are not available at home. Place in the context of a destination marketing mix is a distribution channel for tourism products or services' accessibility (Noo-urai & Jaroenwisan, 2017), which comprises a set of activities undertaken to make the products more accessible and available to target consumers based on location and accessibility (Magatef, 2015). On the other hand, Middleton et al. (2009) suggested that place in tourism is access to a tourism product, confirmation in terms of bookings and reservations, and payments. Furthermore, the distribution of tourism products and services should be followed by promotions, such as trade shows, web pages, resellers, direct messages and travel destinations.

Kotler et al. (2017) stated that the concept of place in a distribution channel from hospitality or a travel distribution system includes travel agents; tour wholesalers; specialists; hotel representatives; national, state and local tourist agencies; consortia and reservation systems; global distribution systems (GDSs); the Internet; and concierges. Bao (2018) revealed that the places where customers buy tourism products can vary widely. Travel agents, tour operators and tour wholesalers are some examples of the tourism distribution channel. The development of the digital era and technological advances have encouraged the establishment of the online travel agent (OTA). It enables tourists to book a travel package such as tourist attractions, accommodation and transportation on a single website. An online travel agent is a website on the World Wide Web dedicated to fulfilling the travel needs of potential tourists. OTAs mainly contain details on travel prices of transportation and accommodation, with various reviews from tourists who have visited these tourist attractions and accommodations.

Indonesia has several OTAs that are quite popular with tourists. Some of them are Traveloka and Tiket.com, which have excellent utility for tourists in independently designing the desired tour. The existence of these two local OTAs provides competition to one of the most popular OTAs in the world—Agoda. This is motivated by the abundant features provided by Traveloka and Tiket.com, which are beneficial to tourists while travelling to tourist attractions, such as the top-up feature, cinema tickets, Internet charging abroad and other features that are not available on Agoda. The price difference between Agoda and Traveloka, and Tiket.com is relatively significant. Agoda displays the price before tax at the beginning of the display, but there is a change if you choose one of the products needed. On the other hand, Traveloka and Tiket.com display the sum of the price included with tax and admin fees.

Promotion

Promotion is one of the essential tools in the marketing mix concept. Rahman et al. (2019) stated that promotion is an activity that explains to the market about the company's products. The essence of promotion in the concept of tourism or destination marketing, as specifically defined by Islam (2020), is that promotion in tourism is an element of a marketing mix intended to inform customers about the products or services and create a brand image about the destination and a position in a competitive market. Besides, the promotion also aids tourism businesses to create positive images. Moreover, Keji and Xuebao (2020) defined the concept of promotion as an activity to create awareness, to maintain, to introduce new products and to increase brand value and service quality. There are various elements of the promotion mix, such as advertising, public relations, personal selling, sales promotion and direct marketing, which are used to communicate customer value and build relationships persuasively

(Kotler et al., 2017). Meanwhile, Islam (2020) added online banners, social media marketing blogs, newsletters or travel promotion videos as elements of the promotion mix.

Indonesia employs branding, advertising and selling (BAS) strategies in promoting its tourism. 'Wonderful Indonesia' and the 'Charm of Indonesia' constitute a branding package carried out by the Ministry of Tourism and Creative Economy as an effort to promote the beauty of tourism in Indonesia to attract visits by foreign and local tourists. Due to the development of the tourism industry 4.0, also known as the digital era, the Ministry of Tourism has been implementing digital marketing as one of the most effective promotional media for disseminating information about tourism in Indonesia. The branding, advertising and selling (BAS) strategy is a promotional strategy carried out by the government followed by a strategy of paid media, owned media, social media and endorser (POSE). The government has been collaborating with various media and several international public figures who have significant influence on Instagram to participate in developing tourism promotion in Indonesia.

People

People play a significant role in service delivery and influence buyers' perceptions. People are very important in the marketing process of any product or service. In the context of the tourism industry, people are not producers, but the products themselves (Bao, 2018). According to Othman et al. (2019), people refer to those involved in service delivery, including those involved in providing service to customers directly or indirectly. Most services require direct interaction between customers and contact personnel, which greatly affects the customers' perception of service quality (Assistant & Sciences, 2015). People cannot be separated from the entire service process. The focal points of people, particularly in the hospitality sector, are behaviour, quality control and personal selling (Rahman et al., 2019). The physical appearance, dressing code, attitude, behaviour, knowledge or competencies, and skills have the strongest impact on customers' perceptions of tourism offers and experiences (Bao, 2018).

People are a key element in a customer-centred company and a way to differentiate variables through products, services, channels and images (Nouri & Soltani, 2015). People are service providers playing a role as the company representatives and are key to the quality of the customer experience. Nowadays, many companies provide specific training to employees, improving interpersonal skills and customer service with a focus on customer satisfaction to ensure that the services provided are of high quality (Bao, 2018).

The Indonesian government and the Ministry of Tourism and Creative Economy have implemented six strategic steps in an effort to accelerate the recovery of the tourism sector and creative economy in Indonesia from COVID-19. The fourth strategy carried out is the onboarding digitalisation programme for creative economy actors. In 2020, this onboarding programme reached 4 million participants, and the Ministry of Tourism and Creative Economy targets this programme to reach 10–15 million tourism and creative economy actors in 2021. This onboarding programme aims at improving capabilities and competencies and not only scaling up in terms of human resources only but also encouraging tourism human resources and the creative economy to produce quality products (Sofia, 2021).

Process

Process involves the actual procedures, mechanisms and flow of service-delivered activities, service delivery and operating systems (Zeithaml et al., 2006). In a tourism context, the process could describe the following activities: moving away from paper-based ordering to digital ordering and account management systems; production of alternative forms of energy, such as sun or wind; supply chain management for food, beverage and workforce; use of low-carbon emission modes of transportation, such as electric or hybrid motorised vehicles to move visitors to, in and from their destinations; and have effective use-minimisation and recycling systems, particularly in environmentally critical locations, such as islands, national parks and protected areas (Pomering et al., 2011).

Process is a system that assists organisations to flow the activities and services delivered to customers and assists customers in making comparisons of different services and products that ultimately contribute to the final purchase decision (Nouri & Soltani, 2015). The process has several indicators, as suggested by Nouri and Soltani (2015), namely government policies in the tourism sector, programmes that suit the needs of tourists and ease of process for buying holiday homes. Ray et al. (2017) argued that process can be measured by personnel participation in serving guests, ease of reservation and government sanction. A different opinion was stated by Yi (2018), who argued that there are several elements of the process, namely service, experience, access, loan/renewal automation, and telephone or online assistance. Another opinion from Soni (2018) stated that processes can be measured by policies, procedures, mechanisation, empiricism, customers, customer direction and flow of activities. Srinivasan (2009) stated that the process in tourism includes (a) planning and anticipating trips, (b) trips to locations/areas, (c) recollection and (d) travel planning packages.

Process in tourism is one of the stages of providing services to tourists. During the COVID-19 pandemic, the tourism ministry carried out countermeasures in all tourist destinations in Indonesia, starting from service quality and promotion to guaranteeing the safety and cleanliness of the tourist attractions. The Ministry issued a cleanliness, health, safety and environmental sustainability (CHSE) programme to support the Indonesian government in implementing the health protocol. Using a mask in every tourist destination, washing hands before entering a tourist destination and physical distancing that applies to every employee and tourist constitute one of the applications of CHSE in each tourist attraction.

Physical Evidence

Service is often intangible, and customers cannot judge that the service is of good quality before they purchase it. Thus, customers use environmental elements as important proxies to evaluate service quality

(Kushwaha & Agrawal, 2015). The service environment is also called servicescape or physical evidence. Physical evidence is a term that describes the real aspects of the product experience, such as physical attributes and facilities (Kwok et al., 2020). In the context of tourism, physical evidence depends on the travel experience, place of residence and comfort (Kwok et al., 2020). The appearance of buildings, landscaping, vehicles, interior furniture, equipment, staff members, uniforms, signs, printed materials and other visible cues serve as physical evidence of a company's service quality (Lovelock & Wirtz, 2010). Physical evidence in tourism is essential both at the individual company level and at the broader destination level (Pomering et al., 2011).

Yi (2018) suggested that there are several elements of physical evidence, namely the environment, buildings, buses, books, lighting and websites. Meanwhile, Soni (2018) stated that physical evidence can be measured by the environment, furnishing, colour, layout, noise level, facilitating goods and tangible clues. Hiransomboon (2012) measured physical evidence, using elements such as having a beautiful building, a beautiful interior decoration, a beautiful surrounding, the required full-scale facilities, a unique building and a decorative character, as well as being clean. Ray et al. (2017) used three indicators to measure the elements of physical evidence, namely rest houses, guides, room service facilities, attractive environments of mediation and rural recreations. Wilson and Bitner (2016) stated that physical evidence comprises servicescape and other tangibles. Servicescape itself comprises exterior facilities (exterior design, signage, parking, landscape and surrounding environment), interior facilities (interior design, equipment, signage, layout, air quality/temperature, lighting, floor coverings and aromas/scents) and virtual environment (web-page design, webcams, photographs, 360-degree views and online chat facilities). Other tangibles include business cards, stationery, billing statements, reports, employee dress, uniforms and brochures.

The Ministry of Tourism and Creative Economy of the Republic of Indonesia seeks to develop tourist destination facilities in Bali to improve the quality of the tourism sector through the Tourism Destination Revitalization Program. There are still several tourist destinations in Bali with inadequate amenities. The main focus of the revitalisation activities is to improve the amenities in many tourist attractions in Bali, in the form of repairing toilets and adding other facilities to support the cleanliness, health, safety and comfort of tourists through the implementation of strict health protocols. The Bali Tourism Destination Revitalization Program was started in early November 2020 in one city and eight districts in Bali. Support for the revitalisation of Bali tourist destinations was adjusted to the priorities and needs of each tourist destination. Cleanliness and health are very important for tourists. Further activities to reactivate the accommodation of health workers in Bali tourist destinations were also carried out. This revitalisation programme is in line with current tourist trends because hygiene and health factors are important in increasing tourist confidence (Zuraya, 2020).

Partnership

One criticism of the service marketing mix model is the absence of a partnership element. Therefore, in the context of tourism, partnerships are an important element that should be included in the marketing mix. Marketers agree that there is a real change in the service industry towards using partnership marketing that is more connected, strategic and effective as a new tool (Salman et al., 2017). Partnerships are about parties working together and brands operating together to reach new customers through a win-win campaign. Marketers agree that there is a real shift in the service industry towards using more connected and strategic partnership marketing as a new, effective tool in the marketing mix. Partnerships are essential not only as a determinant of the success of packaging and programming efforts but also for achieving a holistic systems approach to address the unwanted ecological and sociocultural footprints of tourism. Partnerships can be used to ensure that sustainability resonates along the tourism value chain. The problems of coordination, collaboration and partnership are now at the forefront of much tourism research in finding new solutions to resource management and destination development problems. Destination partnerships are defined as a synergistic relationship between the DMO and other organisations or individuals inside or outside the destination (Sotiriadis & Shen, 2017). The purpose of partnership is to generate consensus and harmony, leading to new opportunities and innovative solutions (Graci, 2013).

Furthermore, Pomering (2017) argued that there are two types of partnerships—internal and external. Internal partnerships depend on internal marketing and human resources management and may include an employee suggestion system. External partners include traditional members of the company's value-delivery network, such as suppliers, channels, customers and other parties that help the company to achieve its goals.

ASEAN Borderless is an idea that was carried out by three tourism ministers in the Association of Southeast Asian Nations (ASEAN), namely the Indonesian Minister of Tourism, Arief Yahya, the Thai Minister of Tourism and Sports, Kopkarn Wattanavrangkul, and the Philippine Minister of Tourism, Wanda Teo, in 2018. They all pushed for the issue of Visit ASEAN @ 50, making Southeast Asia a single destination. The purpose of this idea was to attract foreign tourists to visit ASEAN countries by making ASEAN a single destination. One example of its implementation is in Singapore, which has qualified airlines and transportation management so that Southeast Asia makes Singapore Airlines its regional airlines. On the other hand, Singapore or other countries in ASEAN that are not prominent in their natural and cultural resources can visit destinations in Indonesia to form a mutual relationship (Rakyat, 2017).

Packaging

Packaging is a technique for combining two or more elements of the marketing mix, such as combining different products into a single package. In this case, the term packaging does not mean a new product but rather as a combination of services and products that are interrelated with one another into a package and offering it at a single price, which generally includes resort ticket offers, hotel rooms, flight tickets, destination tickets and any product at a tourist destination (Morrison, 2013). In the tourism industry, packaging is done with the aim of increasing tourist visits by creating price quotes using a special package pricing strategy that is more affordable than purchasing these products separately (Morrison,

2013). From another point of view, packaging can also be interpreted as a collection of tourist desires presented in a complete package to assist the tourists to undertake tourism activities (Vainikka, 2016). In reality, packaging is not only done by the DMO but also travel and tour companies, as well as tour operators, and can be created independently by the tourists themselves.

The packaging is an indicator or benchmark which tourists use to determine the effectiveness of a tour package. Morrison (2013) divided the packaging indicator into two perspectives, consisting of the tourist point of view and the company point of view. From the tourist point of view, five indicators are considered as appropriate benchmarks to measure the effectiveness of the tour package. These indicators are the ability to budget for trips, added value, assurance of consistent quality, convenience and satisfaction. From the point of view of the tourism company, the seven indicators which are considered appropriate benchmarks to measure the effectiveness of the tour package are appeal to specific markets, new target markets, business forecasting, off-peak seasons, customer satisfaction, efficiency and partnering opportunities. Wang et al. (2007) mentioned five indicators in tour packages, which include accommodation, transportation, shopping arrangements, tour leaders and local guides. According to Räikkönen and Honkanen (2013), there are six main elements in tour packages, namely tour operator services, flight services, environment, accommodation services, pre-tour services and airport services. There are different types of tour packages that are based on the special interests of tourists. Some of them are pleasure tourism packages, cultural tourism packages, recreation tourism packages and adventure tourism packages.

Indonesia is one of the emerging economies that has a variety of unique and valuable tourist destinations to be marketed and packaged into attractive tour packages. The Indonesian government and other stakeholders are very supportive of the development of the tourism industry. Various tour packages have been designed by tour operators and travel agents to attract tourists. One of the special tour packages is a halal tour package, which includes various tourist attractions and activities connected with halal, since Indonesia is one of the largest Muslim countries in the world. In 2019, Indonesia was in the first rank of the world halal tourism as a global halal travel destination by the Global Muslim Travel Index (GMTI). This has led to the development of halal tour packages in Indonesia, which is more Muslim friendly by adapting products and services to properly serve Muslim tourists.

Programming

Programming can be defined as the development of a special programme that has various output activities, such as special events and festivals that aim at making tourists spend more in a tourist destination (Morrison, 2013). Programming is an important component of tourism activities in a tourist destination. Programming is included in the packaging of the tourism product to attract potential tourists, and it becomes an added value to be considered in a tour package. The relationship between packaging and programming makes these two terms a strong and most effective strategy for bringing in potential tourists if they are properly designed and actualised.

A programming indicator is a benchmark that is also used to determine the effectiveness and efficiency of a special programme or activity organised during a tour. According to Morrison (2013), programming has eight indicators used as benchmarks, namely participants' experiences, excitement, aesthetics, learning and understanding, trip satisfaction, entertainment, richer experiences and satisfaction of special interest. According to Ayob and Said (2010), four indicators in programming related to a festival or special event include generic features, specific entertainment features, information source and comfort amenities. There are various types of programming in the tourism industry to be considered by potential tourists in choosing the desired type of programmings, such as festivals, special events and special activities.

Currently, the Ministry of Tourism and Creative Economy of the Republic of Indonesia, together with the Directorate General of Immigration, Ministry of Law and Human Rights, is formulating a long-term visa. Unlike the previous visitor visa, the long-term visa has a longer period of five years that can be renewed. The trend of digital nomads is currently rife all over the world. One of the causes is the COVID-19 pandemic, which has changed people's lifestyles. Those who initially had to work in an office became free without being limited by space and time. This trend is considered very appropriate to be applied in Bali, considering that Bali has natural beauty and is equipped with adequate telecommunications infrastructure. This long-term visa is one of the main preconditions so that more digital nomads in the world will consider Bali as their second home. Besides, more people are now working from home. Bali offers a healthy lifestyle, properties that are in good condition and more affordable culinary delights. With so many foreign tourists making Bali their 'second home', the property business in Bali is expected to grow (Khairunnisa, 2021). This programme provides tourists with a unique opportunity that urban Bali will never experience. Tourists can help villagers while on vacation in Bali and get to know the deeper Balinese culture and heritage.

Conclusion

The marketing mix is a series of strategies employed by marketers to fulfil the needs of their target markets. Each element in the marketing mix is a series that has continuity with one another. One of the marks of success of the tourist destination is its ability to formulate an optimal marketing mix policy to support the tourism activities and attract potential tourists to visit the country. In this case, the Indonesian Ministry of Tourism and Creative Economy, which is responsible for formulating tourism policies, has developed a marketing mix strategy to assure the success of all the programmes designed. How the Indonesian government matches each element in the marketing mix with its target market will lead to a strong brand positioning. This strategy could be adopted by other emerging tourist destinations.

References

Amzad, M., Sarker, H., Candidate, P., Aimin, W., & Begum, S. (2012). Investigating the impact of marketing mix elements on tourists 'satisfaction: An empirical study on east lake. *European Journal of Business and Management, 4*(7), 2222–2839.

Assistant, M., & Sciences, F. (2015). The Impact of tourism marketing mix elements on the satisfaction of inbound tourists to Jordan head of marketing department. *International Journal of Business and Social Science, 6*(7), 41–58.

Avlonitis, G. J., & Indounas, K. A. (2007). An empirical examination of the pricing policies and their antecedents in the services sector. *European Journal of Marketing, 41*(7/8), 740–764. https://doi.org/10.1108/03090560710752384

Ayob, N., & Said, A. (2010). Service quality and customer satisfaction within festival and special event. In *Knowledge Management International Conference (Kmice2010), May* (pp. 32–37).

Bao, H. (2018). Marketing of tourism services/experiences. In *The Emerald handbook of entrepreneurship in tourism, travel and hospitality* (pp. 261–275). https://doi.org/10.1108/978-1-78743-529-220181021

Benjamin, B. (2010). Promoting an Emerging tourism destination. *Global Journal of Management and Business Research, 10*(1), 21–28.

Booms, B., & Bitner, M. (1981). Marketing strategies and organisation structures for service firms. In *Marketing of Services: 1981 Special Educators*. American Marketing Association.

Buhalis, D., & Costa, C. (Eds.). (2005). *Tourism business frontiers*. Heinemann.

Ciriković, E. (2014). Marketing mix in tourism. *Academic Journal of Interdisciplinary Studies, 3*(2), 111–116. https://doi.org/10.5901/ajis.2014.v3n2p111

Dwyer, L., & Forsyth, P. (2011). Methods of estimating destination price competitiveness: A case of horses for courses? *Current Issues in Tourism*, 751–777. https://doi.org/10.1080/13683500.2011.586680

Fyall, A., & Garrod, B. (2005). *Tourism marketing. A collaborative approach*. Channel View Publications.

Graci, S. (2013). Collaboration and partnership development for sustainable tourism. *Tourism Geographies, 15*(1), 25–42. https://doi.org/10.1080/14616688.2012.675513

Henderson, J. C. (2015). The new dynamics of tourism in South East Asia: Economic development, political change and destination competitiveness. *Tourism Recreation Research, 40*(3), 379–390.

Hiransomboon, K. (2012). Marketing mix affecting accommodation service buying decisions of backpacker tourist traveling at inner Rattanakosin Island in Bangkok, Thailand. *Procedia Economics and Finance, 3*(12), 276–283. https://doi.org/10.1016/s2212-5671(12)00152-9

Islam, M. R. (2020). Tourism marketing dimension on Bangladesh perspective. *Asian Journal of Social Science and Management Technology, 2*(6), 16–25.

Keji, D., & Xuebao, D. (2020). *Linkage between sustainability and tourism marketing mix strategy*. https://doi.org/10.37896/jxu14.4/145

Khairunnisa, S. N. (2021). *Nomad staycation di Bali, kemenparekraf rumuskan long term visa untuk turis asing*. https://travel.kompas.com/read/2021/03/17/194148527/nomad-staycation-di-bali-kemenparekraf-rumuskan-long-term-visa-untuk-turis

Kotler, P., Bowen, J. T., Makens, J. C., Baloglu, S., & Edition, G. (2017). *Marketing for Hospitality and tourism* (7th ed.). Pearson Education Limited.

Kotler, P., & Keller, K. (2006). *Marketing management* (12th ed.). Prentice Hall.

Kushwaha, G. S., & Agrawal, S. R. (2015). An Indian customer surrounding 7P[U+05F3]s of service marketing. *Journal of Retailing and Consumer Services, 22*, 85–95. https://doi.org/10.1016/j.jretconser.2014.10.006

Kwok, L., Tang, Y., & Yu, B. (2020). The 7 Ps marketing mix of home-sharing services: Mining travelers' online reviews on Airbnb. *International Journal of Hospitality Management, 90*(August 2019), 102616. https://doi.org/10.1016/j.ijhm.2020.102616

Lovelock, C., & Wirtz, J. (2010). *Service marketing*. Pearson Education.

Magatef, D. S. (2015). The impact of tourism marketing mix elements on the satisfaction of inbound tourists to Jordan. *International Journal of Business and Social Science, 6*(7), 41–58.

McCarthy, E. J. (1960). *Basic marketing: A managerial approach*. R. D. Irwin.

Middleton, V. T. C., Fyall, A., Morgan, M., & Ranchhod, A. (2009). *Marketing in travel and tourism* (4th ed.). Elsevier.

Morrison, A. M. (2009). *Hospitality and travel marketing* (4th ed.). Cengage Learning.

Morrison, A. M. (2013). Marketing and managing tourism destinations. In *Marketing and managing Tourism Destinations*. https://doi.org/10.4324/9780203081976

Noo-urai, N., & Jaroenwisan, K. (2017). The sustainability marketing mix for Thai senior tourism. *International Journal of Social Sciences and Management, 4*(2), 80–89. https://doi.org/10.3126/ijssm.v4i2.17157

Nouri, B. A., & Soltani, M. (2015). Evaluating the effect of tourism marketing mix on buying holiday homes in Cyprus. *International Journal of Business Administration, 6*(5), 63–74. https://doi.org/10.5430/ijba.v6n5p63

Othman, B. A., Harun, A., Rashid, W. N., Nazeer, S., Kassim, A. W. M., & Kadhim, K. G. (2019). The influences of service marketing mix on customer loyalty towards umrah travel agents: Evidence from Malaysia. *Management Science Letters, 9*(6), 865–876. https://doi.org/10.5267/j.msl.2019.3.002

Pomering, A. (2017). Marketing for sustainability: Extending the conceptualisation of the marketing mix to drive value for individuals and society at large. *Australasian Marketing Journal, 25*(2), 157–165. https://doi.org/10.1016/j.ausmj.2017.04.011

Pomering, A., Noble, G., & Johnson, L. W. (2011). Conceptualising a contemporary marketing mix for sustainable tourism. *Journal of Sustainable Tourism, 19*(8), 953–969. https://doi.org/10.1080/09669582.2011.584625

Rahman, M., Islam, M. S., Al Amin, M., Sultana, R., & Talukder, M. I. (2019). Effective factors of service marketing mix on tourist satisfaction: A case study. *Asian Social Science, 15*(7), 1–15. https://doi.org/10.5539/ass.v15n7p1

Räikkönen, J., & Honkanen, A. (2013). Does satisfaction with package tours lead to successful vacation experiences? *Journal of Destination Marketing and Management, 2*(2), 108–117. https://doi.org/10.1016/j.jdmm.2013.03.002

Rakyat, P. (2017). *ASEAN Borderless: Kerjasama di Bidang Pariwisata*. https://www.pikiran-rakyat.com/wisata/pr-01279286/arief-yahya-gulirkan-ide-aseansebagai-destinasi-tunggal-pariwisata-400023

Ray, N., Das, D. K., & Kumar, R. (Eds.). (2017). *Tourism marketing: A strategic approach*. CRC Press. https://doi.org/10.1201/9781315365862

Salman, D., Tawfik, Y., Samy, M., & Artal-Tur, A. (2017). A new marketing mix model to rescue the hospitality industry: Evidence from Egypt after the Arab Spring. *Future Business Journal, 3*(1), 47–69. https://doi.org/10.1016/j.fbj.2017.01.004

Sofia, H. (2021). *Menjelang dibukanya kembali dunia pariwisata Indonesia*. https://www.antaranews.com/berita/2037266/menjelang-dibukanya-kembali-dunia-pariwisata-indonesia

Soni, G. (2018). A study of extended p's (people, process and physical evidence) of marketing mix in wildlife tourism with special reference To Kanha National Park. *International Journal of Current Issues in Life Sciences, 7*(5), 2131–2137.

Sotiriadis, M., & Shen, S. (2017). The contribution of partnership and branding to destination management in a globalized context: The case of the

UNWTO Silk Road Programme. *Journal of Tourism, Heritage & Services Marketing, 3*(2), 8–16.

Srinivasan, K. (2009). Tourism marketing: A service marketing perspective. *Munich Personal RePEc Archive, 14031*, 1–6. https://mpra.ub.uni-muenchen.de/14031/1/MPRA_paper_14031.pdf

Vainikka, V. E. (2016). Tourist guide reflections on the spatialities of mass tourism: A case study of Finnish package tourism in Crete. *Fennia, 194*(1), 64–78. https://doi.org/10.11143/40925

Wang, K. C., Hsieh, A. T., Chou, S. H., & Lin, Y. S. (2007). GPTCCC: An instrument for measuring group package tour service. *Tourism Management, 28*(2), 361–376. https://doi.org/10.1016/j.tourman.2006.04.017

Wilson, Z., & Bitner, G. (2016). *Services marketing: Integrating customer focus across the firm* (European edn.). McGraw-Hill.

World Economic Forum. (WEF). (2012). *The ASEAN travel & tourism competitiveness report 2012*. WEF.

Yi, Z. (2018). *Marketing services and resources in information organizations*. Chandos. https://doi.org/10.1016/b978-0-08-100798-3.00005-2

Zeithaml, V., Bitner, M. J., & Gremler, D. D. (2006). *Services marketing, integrating customer focus across the firm*. McGraw-Hill.

Zuraya, N. (2020). *Kemenparekraf revitalisasi destinasi wisata di Bali*. www.republika.co.id

3

Special Events and Destination Marketing Strategies in Emerging Economies in Asia

Mohd Raziff Jamaluddin and Wan Soraya Wan Abdul Ghani

Introduction

Special events are regarded as a part of the travel product and have been considered vital to the tourist sector, as they add value to the destination. Meanwhile, sporting events account for a sizable portion of global tourism and have developed into a powerfully dynamic niche market, as enthusiasm and involvement in sports have increased in recent years and gained popularity. Destination management organizations have been engaging in special events to brand and reinforce the local image. Thus, defining the event's position in destination marketing would assist destination managers in developing strategies to encourage tourism growth and strengthen the destination's

M. R. Jamaluddin • W. S. W. A. Ghani (✉)
Universiti Teknologi MARA, Selangor, Malaysia
e-mail: raziff@uitm.edu.my; wansor2956@uitm.edu.my

I. Mensah et al. (eds.), *Marketing Tourist Destinations in Emerging Economies*, Palgrave Studies of Marketing in Emerging Economies,
https://doi.org/10.1007/978-3-030-83711-2_3

competitive edge. This chapter delves further into the evolution of destination marketing strategies, including the roles of destination marketing organizations (DMOs) and case studies from emerging economies. The first part of the chapter covers destination marketing that looks into the growth in the Association of Southeast Asian Nations (ASEAN) and a case study of Tourism Malaysia, the DMO for Malaysia. The chapter then examines the importance of special events as tourism growth catalysts. The three types of special events namely local events, major events and mega-events are discussed. The chapter then focuses on sporting events by taking two case studies, specifically, the Republic of Korea and Malaysia.

Destination Marketing Strategies

Sotiriadis (2021, p. 42) defines destination marketing as "a continuous, sequential process through which a DMO plans, researches, implements, controls and evaluates programs aimed at satisfying tourists' needs and wants as well as the destination's and DMO's visions, goals and objectives". According to Kruja (2018), destination marketing strategy is a common research subject because it provides a variety of opportunities, like (a) the development and implementation of research programmes that track, evaluate and clarify key market trends and conditions; (b) guide performance reporting programmes that provide reliable and effective programme assessments; (c) the management of a communication programme; (d) DMOs will serve as a point of contact for government officials and tourism industry leaders in the creation of cooperative marketing programmes. Foris et al. (2020) noted rapid shifts in tourism demand and consumption, intensified competition and the advent of new tourist destinations. They added that the implications of information technology development necessitate a new approach to implementing modern management techniques to ensure effective destination marketing strategies.

The Association of Southeast Asian Nations (ASEAN, 2017) strives to place and brand the region as a single destination, reducing consumer uncertainty and ensuring clarity. As the plan progresses, the challenge for ASEAN is to promote subregional destinations and routes with more inclusive and measurable results. ASEAN (2017) has identified a number of threats, including global trends and forces such as:

- Tourism growth
- Technological advancements
- Sharing economy
- Rapid urbanization
- Increasing uncertainty
- Changing visitor expectations
- Demographic changes

The most critical force, as noted by ASEAN (2017), is to meet changing visitor expectations because visitors are able to recognize and demand constantly evolving and emerging experiences, such as safety and security, and culinary tourism. Furthermore, the Organisation for Economic Co-operation and Development (OECD, 2018) projected that the visitor economy will change as income and education levels in developing markets increase, gender imbalances are addressed, the global population continues to age and new consumer groups emerge. Consequently, Millennials and Generation Z will become important drivers of mobility, and their travel habits will cause significant changes in the tourism industry.

Destination Marketing Strategies: Tourism Malaysia

From the first blueprint in 1966 to 2015, Tourism Malaysia's marketing and promotion strategies focused on increasing tourist arrivals in the short term and establishing a strong image in the long term (Hussin & Buchmann, 2019). It is worth noting that the grand launch of Visit Malaysia Year 2014 coincided with the launch of a digital marketing campaign that centred on leveraging digital assets to maximize presence in the digital sphere, growing and sustaining a good relationship with fans and followers on social media, and eventually turning them into real tourists or repeat visitors (Tourism Malaysia, 2014). As reported by the Standing Committee for Economic and Commercial Cooperation of the Organisation of the Islamic Cooperation [COMCEC] (2015), Tourism Malaysia discovered that with shifts in traveller behaviour as a result of advances in digital technology, including prevalent mobile communication, 24/7 communication activities, the power of visual information, travellers' gain from search engine advancements, common location-based services, and with the abundance of causes, platforms and media influencing the decision-making phase, the traveller experience has become more complex.

Digital marketing platforms used by Tourism Malaysia include a YouTube channel, web TV, social media, Tourism Malaysia website and an official mobile travel application. It was reported that the digital marketing campaign programme had recorded over 12.5 million views (Tourism Malaysia, 2014). The shifts in traveller behaviour require DMOs to make use of mobile application features. Tourism Malaysia used the "Trip Planner Mobile App" to provide travellers with personalized, inspiring and practical information about Malaysia, including things they could do while in the country. The most important aspect is that it allows travellers to prepare an itinerary that includes information about tourism sites.

Websites have long been a common feature of destination marketing. The website for Tourism Malaysia presents a vivid and colourful picture of Malaysia in 16 languages. Tourism Malaysia has interactive websites, which use a lot of visual presentation, information content and communication tools to enable visitors to explore the available information. The picture of nature on Tourism Malaysia's website is fascinating and interesting. According to Hamid et al. (2017), this indirectly influences people's decision to travel to Malaysia. The study also discovered that the photo gallery's colour and design make it easy to visualize the destination.

Interestingly, Tourism Malaysia (2016) launched a sales campaign at the beginning, middle and end of the year to promote shopping tourism in Malaysia. In 2016, a social media icon named "Miss SHOPia" was created to help Tourism Malaysia develop a stronger online social media presence. The icon has grown into the ultimate shopping buddy, guiding consumers on the latest fashion trends, when and where to shop, what to watch out for and how to get the most out of their shopping experience in Malaysia (Tourism Malaysia, 2016). Miss SHOPia was used in advertising materials such as street buntings, posters, t-shirts, accessories, luggage tags and notebooks, in addition to having her own website. According to Heras-Pedrosa et al. (2020), the destination should focus on the brand formation and promotion by demonstrating, persuading and attracting visitors to the destination.

Special Events

People travel thousands of miles to meet up with friends and colleagues for face-to-face meetings and other events. The importance of meetings and events makes like-minded individuals meet, learn and get involved in formal and informal activities. As a result, the mass gathering of people for events has advanced from a merely societal concept to a business-related concept. The idea of hosting events has changed the way destination marketing organizations (DMOs) coordinate and promote the events industry as one of the sub-sectors in the tourism industry. From a social perspective, events enhance the awareness of the destination brand image, foster cultural exchanges and promote local culture preservation (Wan & Jin, 2019). As mentioned by Getz (1989), special events are a distinct form of tourist attraction that ranges in size from mega-events like the Olympic Games and the World Fair to smaller-scale events that are held at parks and facilities.

Consequently, Getz (2008) described a special event as a spatial-temporal phenomenon, with each one being distinct due to interactions between the environment, individuals and management systems, including design elements and the programme. Nguyen (2017) mentioned that special events are unique events that can be considered a significant component of the tourism industry and viewed as an industry by local governments that generate them. Getz (2008) noted that the main element

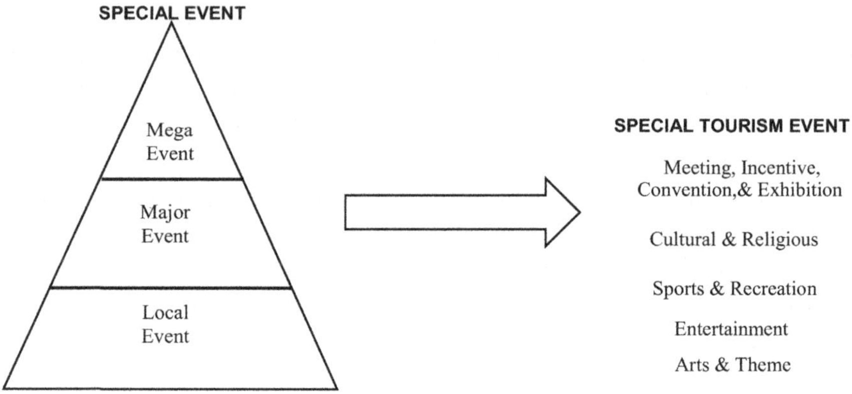

Fig. 3.1 The progress of mass tourism events

of a special event is "be there to truly enjoy the experience". He further elaborated that from the event organizer's perspective, special events are:

- One-time or infrequently events that are organized outside of the sponsoring or coordinating body's daily programmes or activities.
- An occasion for leisure, social or cultural activity outside of the usual range of options or outside the daily experience of the customer or visitor.

As shown in Fig. 3.1, special events are classified based on the number of attendees, the categories of participants and the extent to which they are covered.

Local Events

Looking at the pyramid of events, the local event, which caters to domestic tourists, is the first and perhaps the smallest-scale event. It is usually organized by the local event managers with the help of the local community. According to Mihajlović (2017), the primary goal of a local event is to appeal to the entertainment needs of the local community. These activities do not only help the economy but also raise local pride and self-awareness, foster tolerance and facilitate participation in sports and cultural events. Mihajlović added that local activities were typically planned as part of the local government's mission to foster social and cultural diversity. They are often held in public places such as parks, streets and schools, and they are organized by a variety of community organizations with the help of volunteers. This form of event generally attracts a small number of foreign tourists who came to engage in cultural activities. Malaysia, as one of the emerging economies, is recognized as a multicultural country that has established itself as Asia's tourism hub (Tourism Malaysia, 2016). Culture and creative economy have developed rapidly across the world in the twenty-first century in response to demand. This rise in demand can be attributed to the increasing middle class and disposable income around the world.

Since 1999, Tourism Malaysia, Malaysia's DMO, has used the "Malaysia Truly Asia" campaign to celebrate Malaysia's diversity and to offer diversity in the highly competitive Southeast Asian tourism market, which is primarily centred on beaches, shopping and nightlife (ISC, 2020). At the heart of the "Truly Asia" campaign are the multi-ethnic rituals of traditions, beliefs and religions. Malaysia has taken advantage of its festivals and events to diversify its tourism products. Colours of Malaysia, also known as *Citrawarna* by locals, has been one of Malaysia's most important cultural festivals since 1999, showcasing the country's rich customs and culture, traditional dance and music, arts and crafts, and a wide range of local foods. *Citrawarna*, which has been catering to domestic tourists since 1987, has been revitalized, thanks to the "Malaysia Truly Asia" campaign. In 2012, *Citrawarna* was patronized by more than 20% of the foreign tourists who visited Malaysia to experience the multicultural performances and the unique culture of the country. Aside from that, more than 60% of the visitors were under the age of 35 years and had an annual income of less than US$1000. The survey also discovered that locals rely on television for information, while foreign visitors rely on brochures and posters as sources of information. According to Tourism Malaysia (2016), different themes are used as branding strategies for *Citrawarna*, with the idea of "Myth and Folklore" and the theme "Unleash the Moment, Unfold the Stories" used in 2016 to highlight the various folklores of Malaysian states.

Major Events

The major events, on the other hand, are more appealing to international audiences and cater to niche markets. A major event is typically coordinated by local event organizers with the involvement of various local agencies. Oklobdžija (2015) defined a major event as a large-scale event that generates strong public attention and media coverage. According to him, a major event aims to attract a large number of visitors, especially those who are sports-oriented, with an international reputation and a clearly established competition structure. Mihajlović (2017) noted that major events aim to raise tourists' interest in a particular destination or area. Andersen et al. (2015) suggested that sporting events are typically treated as major events because they are coordinated in the city centre arena and have a new perspective on how to view the destination as regular leisure activities. However, Oklobdžija claimed that a major event could harm the destination's reputation because tourists will be more interested in the event than in the destination itself, which will have no long-term effects, and a major event will not be able to attract enough tourists if the destination is unknown.

Malaysia Major Events (MME), a division of Malaysia Convention and Exhibition Bureau (MyCEB), was formed to identify, promote, encourage and support the staging of viable international events in Malaysia (MyCEB, 2015). With the primary goal of boosting Malaysia's economic growth and profile, MME is also tasked with identifying and supporting major event bids in sports, arts, lifestyles and entertainment. The division also assists home-grown and home-hosted events, in order to further improve Malaysia's global appeal as the region's destination of choice for major events. Though MME's most important marketing strategy is event bidding, it also engages in other marketing activities such as advertisement and promotional campaigns, which have also played a significant role in establishing Malaysia as a major event destination. For instance, MME advertised in industry magazines for worldwide coverage and also ran the "Malaysia Like Never Before" campaign in the Meetings, Incentives, Conventions and Exhibitions (MICE) market (MyCEB, 2018). The production of sales and marketing collateral is another important aspect of brand marketing, with exhibition sales info kits/fact-sheets and MyCEB info kits being distributed at roadshows and networking events. As a result, ten international events with nearly 20,000 international participants were secured in 2018, with a cumulative projected tourist expenditure of US$6.4 million (MyCEB, 2018).

Mega-Events

Mihajlović (2017) suggested that mega-events are planned events that have a national economic effect and attract a large amount of media coverage. Müller (2015) proposed that mega-events are ambulatory, fixed-duration events that draw (a) a large number of visitors, (b) have a large mediated range, (c) have high costs and (d) have major impacts on the built environment and community. As noted by Kim and Kang (2020), mega-events are projected to benefit host cities in a number of ways, including increased city branding, economic benefits, non-infrastructure benefits and revenue from ticket sales and tourism. Kim and Kang added that profits from tourism (resulting from a better city image) are the most highly anticipated contributions of mega-events from which local communities will benefit. In emerging economies, the government offers financial grants to the mega-event organizers through the tourism ministry and the regional government in order to speed up the development of home-grown mega-events.

Again, the Malaysia Convention and Exhibition Bureau (MyCEB) plays a key role in promoting mega-events. MyCEB was given a critical role in positioning Malaysia as a leading mega business events destination under the Entry Point Project (EPP) 10 of the Tourism National Key Economic Area in the nation's Economic Transformation Programme (ETP), which was launched in 2010 (MyCEB, 2020). According to the report, MYR3.9 billion in additional gross national income, 16,720 new jobs and 2.9 million business event visitors were among the major benefits of the mega-events business in Malaysia. Thus, recognizing the competitive environment is one of MyCEB's strategies for achieving the status of a mega-event hub in Malaysia.

Sporting Events as Special Events

Sporting events have become an important segment of the tourism industry. It has become one of the fastest-growing segments of the tourism industry. Through a cooperation agreement formed in 2001, the World Tourism Organization (WTO) and the International Olympic Committee (IOC) recognized sports and tourism as important for mutual understanding, thereby bringing considerable attention to the sector. According to Pitts and Stotlar (2013), the sports industry comprises individuals, companies and organizations interested in developing, supporting, promoting or organizing any sport-related activity, experience or business venture. The sports industry has also produced segments such as leisure, wellness and betting. Sporting activities have long been recognized as an important component of sports tourism, and hosting one can provide a destination with additional benefits due to their popularity (Khor et al., 2019). Along with this growth, destination marketers are paying special attention to the strategy of trying to distinguish the destination's image and gain international recognition through hosting sporting events (Wafi et al., 2018). The sporting events have attracted the interest of scholars who want to take a more systematic approach to research on the issue. It is backed by well-developed tourism literature that dates back to the early 1990s and has expanded exponentially since 2000 (Getz, 2012).

Sports tourism events offer an exciting opportunity to brand cities as preferred tourist destinations and to provide growth prospects for local economies due to the fact that sport tourism events play a critical role in defining national and local tourism offerings, as well as converting destinations into attractive event-driven economies (Knott & Hemmonsbey, 2015; Siyabulela, 2016). There have been criticisms that organizing sporting events at the international level is a waste of money because the financial requirements of hosting such events are immense (Mapjabil et al., 2017). Sporting events are typically held in various locations, resulting in varying levels of tourist activity during the event. Mapjabil et al. (2017) suggested five dimensions of sporting events as follows:

- The duration of the event
- Preparations before the tournament
- Media coverage
- Sponsorship
- Individuals involved

Sporting events, such as the Olympic Games, World Cup, Commonwealth Games, NFL Super Bowl, Grand Slam and Formula One Grand Prix, are the most popular sporting events globally. In contrast to other special events, sporting events attract the most media attention. The event usually lasts for more than a week and, in some cases, the entire calendar year. For example, the Olympic Games lasts for more than two weeks, while the World Cup lasts for one month. The hosting city has to spend a lot of time and money to prepare for the events, so it is necessary to host them over a long period of time. The duration of the event helps the hosting city obtain a return on investment in the form of a short-term increase in economic activities and trade, especially in the course of the event. Besides, improvements in facilities and infrastructure and the addition of more hotels and resorts tend to have a long-term effect on the tourism sector in the hosting city.

Mapjabil et al. (2017) noted that before a sporting event takes place in a given state or region, the host may make detailed arrangements in terms of sports facilities, accommodation for participants, venue maintenance, transportation and a variety of other activities. To ensure the welfare of the athletes and delegates, the host must also provide support personnel, including volunteers and equipment operators. A site suitability assessment should be conducted at the venue, including ground conditions, traffic and pedestrian paths, topography and the location of available facilities, such as water, electricity and audience capacity (Kent County Council, 2016). Furthermore, Kent County Council (2016), among others, suggested that (a) clear communication lines should be established and maintained throughout the planning stage and during the event; (b) all bookings, decisions and transactions prior to the event should be put into writing to ensure clarity and to provide a record; and (c) targets and deadlines for the period leading up to the event should be set to ensure the key stages are not missed.

Mapjabil et al. (2017) have added media attention as another dimension of sporting events. Previously, print and electronic media were the media through which events were marketed. A creative company was commissioned to run the coverage and assess the effectiveness of the advertising programme for the sporting event. While social networking is a common tool for engaging with followers, there is no real strategy for using social media to engage sports fans. As Nufer (2016) points out, there is no real evidence of past sporting events that used social media for event promotion and participation, despite many sponsored brands attempting to share views with fans. Furthermore, due to the small number of spectators at the stadia and the sports industry's lack of marketing enthusiasm, attracting fans through social media became a secondary focus (Sato et al., 2018). However, social media will be used as the strategic communication tool that is able to profile followers for future marketing purposes. Most notably, digitalization has made it possible for sports fans to access content outside of venues (Deloitte, 2020).

Sponsorship is the most significant consideration in the execution of a sports event. As Mapjabil et al. (2017) point out, in the concrete, obtaining sponsorship for an organization enables the organizer to prepare a financially viable event. Official sponsors, such as television networks, beverage manufacturers, airlines, telecommunications companies, automobile companies, energy products and household brands, can generally spend a large amount of money on sponsorship for branding purposes. Sponsorship in sporting events is unique, as indicated by Deloitte (2020). He pointed to core aspects of sports sponsorship, and one of them is the effect of recent technological breakthroughs on the industry.

Additionally, Mapjabil et al. (2017) proposed individuals involved in sporting events as the final dimension. Athletes, team managers, coaches, physiotherapists, team officials and other delegates are all engaged in sporting events. As demonstrated by the Olympic Games, the overall number of people participating may be tens of thousands. The sports competitors (athletes and delegates) and spectators are the two main groups of people who attend sporting events (Hover et al., 2016). Sports participants from the host city are critical for promoting social impact in the form of indirect benefits of the sporting event. Individuals involved are essential for the host destination to gain in terms of expenses and post-event tourism, as the probability of spreading positive word of mouth is the most lucrative long-term effect of sporting events (Lintumäki et al., 2020).

Sporting Event as a Catalyst for Special Events in Asia

From 2018 to 2022, the Asian continent has been the focal point of sporting events. The 2018 Winter Olympics in Pyeongchang marked the start of the Asian continent's busiest time of hosting sporting events. With two mega-events, the 2018 Winter Olympic Games and the 2019 World Swimming Championships, the Republic of Korea has established itself as one of Asia's most successful hosts of sporting events. Similarly, Japan is Asia's most active host of sporting events, organizing a number of sporting events in 2019. In 2019, Japan hosted three major sporting events: the Rugby World Cup 2019 (RWC2019), the 24th Women's Handball Championship and the Volleyball World Cup. The ability of Japan to host three major sporting events in the same calendar year resulted in huge cash inflows, bolstering Japan's position as one of the world's most powerful economies.

The Rugby World Cup 2019 (also known as RWC2019) brought in more than US$3 billion in revenue to the 12 cities that hosted the tournament. Approximately 400,000 foreign spectators visited Japan during the tournament and spent an average of US$4000 during their stay (Asia Outlook, 2019). The high cost of living, as well as exorbitant entrance tickets and hotel rates, could not deter fans from coming to watch the game. During the RWC2019, Japan generated over 25,000 jobs with just a modest investment in infrastructure (Asia Outlook, 2019).

In 2020, Japan was to host the second most prominent sporting event in the world, the summer Olympic games. Tokyo Olympic 2020 was going to be Japan's second time hosting the Olympics. Tokyo had undergone major infrastructure development that laid the foundations of modern Japan since the 1964 Olympic games. The Olympic 2020 was the most long-awaited sporting event of the year due to Japan's ability to exhibit groundbreaking displays and advanced technologies. However, due to the COVID-19 pandemic, the event was rescheduled to 2021.

Apart from Japan, China has also played an important role in the hosting of sporting events. Beijing Olympic 2008 illustrated China's emergence as a global economic force, as well as shifting people's perceptions of China's old political views (China Business Review, 2008). China was depicted as a peaceful, civilized, harmonious and modern nation at the Beijing Olympics. China spent nearly US$40 billion on infrastructure, paving the way for other sectors such as advertising, media, movies, the Internet, mobile and portable devices, renewable energy, sports and tourism (China Business Review, 2008). Beijing will also host the Winter Olympics for the first time in 14 years in 2022.

In 2022, Qatar, a small Arabian Gulf nation with a population of 2.8 million people, will host the world's largest sporting event, the football World Cup. The World Cup will be held in the Middle East for the first time in history. Qatar, which is 828 times smaller than China, will host more than 500,000 football fans worldwide. The development of facilities is still ongoing, with eight state-of-the-art stadia set to open soon. A total of 64 football matches will be played throughout the month-long event.

Sporting Events: A Case of the Republic of Korea

Sporting events offer major benefits to the host city, as exemplified by The Republic of Korea, which is popularly known as South Korea. South Korea was founded in 1945, following the end of World War II. The country was governed by the military after achieving independence in 1948, before transitioning to a full liberal democracy in the 1980s. Due to social and political instability, South Korea was one of Asia's poorest countries at the time. In the 1970s, South Korea started to prioritize heavy industries and a strategy of export-based industrialization. To support its manufactured products, the country began to establish diplomatic relations with more developed nations.

The year 1988 marked the start of South Korea's transformation into a modern country. South Korea, a third-world country with a land area of 100,210 km^2, was the first third-world country to host the world's largest sporting event, the Olympic Games, in 1988. The event took place on a 420-acre land in the heart of the city. The construction of an integrated city that connected the Olympic venues and the airport through highways and metro trains was completed. High-rise office buildings, homes, housing areas and digital networks have all been designed in more urban areas (International Olympic Committee, 2018). Seoul's image had changed dramatically since 1988, when it was transformed from a quiet administrative capital to a booming metropolis with a population of over 10 million people.

In comparison to Japanese products, Korean household brands at the time lacked performance. The Olympic games of 1988 offered a platform for brands such as Daewoo, GoldStar, Pro-Specs and Lotte to develop their brands. The goods were able to break into the foreign market due to sponsorship and licensing. Seoul Olympic 1988 raised over US$3.3 billion in sales and provided 20,744 new jobs in terms of economic impact (International Olympic Committee, 2018). The Seoul Olympic Games of 1988 had a huge impact on South Korea. The automobile industry was also improved as a result of facilitating movement within cities. The creative and entertainment industries benefited as well. Many agencies were established as a result of marketing, sponsorship and licensing activities, which recruited local celebrities and talents, resulting in the formation of Korean Pop Culture. Furthermore, in 1988, South Korea's annual trade exceeded US$100 billion for the first time. Before the Seoul Olympic Games in 1988, South Korea was a developing country. In reality, South Korea's GDP was comparable to that of Malaysia in the 1970s. South Korea's GDP per capita rose significantly in the year leading up to the 1988 Olympic Games (1985–87) and was two times higher than Malaysia's GDP per capita after the Games.

The Olympic Games in Seoul in 1998 were a crucial factor in South Korea's tourism growth. The high-speed rail system that linked Korea's major cities boosted domestic tourism. Following the Olympics, many people travelled to Seoul to see the event venue as well as the modern city. Furthermore, the upgraded airports resulted in improved connectivity and frequency of flights to Seoul. The high frequency of outbound flights operated by Korean Air and Asiana Air resulted in an increase in outbound tourism. From a mere 1 million tourists in 1980 to over 3 million tourists in 1990, the number of visitors increased by threefold.

Other tourism sub-industries arose as a result of Seoul's rapid post-Olympic growth. Korean culture and historical values were actively promoted as tourism products. The MICE industries were also created, and Seoul Metropolitan is now noted for such events. Tourist arrivals figures had been increasing year after year before the outbreak of COVID-19, with 2019 recording a 14% rise from 15.3 million to 17.5 million visitors (Statista, 2020).

Sporting Events: A Case of Malaysia

Malaysia was inspired to host the 1998 Commonwealth Games after the success of the Seoul Olympic Games in 1988. The Commonwealth Games was held in Asia for the first time in history. The Games served as a catalyst for infrastructure development in Kuala Lumpur, Malaysia's capital. The new airport was established on the outskirts of the city as part of a long-term strategy to expand the area's satellite cities. Highways and light rail transport were used to improve the link between major urban areas and the event site. With the main event arena that can handle up to 85,000 fans, the purpose-built sporting complex became the focal point of the games.

With the inaugural Le Tour De Langkawi in 1996, Malaysia's hosting of international sporting events began a few years before the 1998 Commonwealth Games. Since then, Langkawi has been the epicentre of sporting events. The Langkawi Development Authority (LADA), a specialized government agency charged with planning Langkawi's development as a tourist destination, developed the proposal to host the international sporting event. Because of the minimal impact of tourism on the island, sports tourism has become the primary product for attracting foreign visitors. Langkawi is surrounded by 99 islands in the Andaman Sea. Langkawi's geographical features, including scenic views and a lush green backdrop, make it an ideal venue for sporting activities. Almost every month, sporting events centred on the beach and outdoor sports are held. In order to maintain Langkawi's natural beauty, sporting events are also favoured as tourism products.

The major sporting events held in Langkawi had a robust economic effect on the island's tourism industry. The International Federation Muay Thai Championship 2014 (IFMA2015), for example, was the first to be held outside Thailand. The event drew a more significant number of foreign tourists than domestic visitors. Foreigners made up about 52% of the attendees, who spent an average of nine days in Langkawi. During their stay in Malaysia, the bulk of the foreign attendees spent around US$2400. The foreign attendees set aside half of their total budget for their stay in Langkawi, and the other half was spent in Kuala Lumpur before returning to their home countries.

Ironman Langkawi is the most popular sporting event in Langkawi. Each version of the triathlon has drawn over 1000 participants. The most common age group at Ironman Langkawi is those between the ages of 26 and 40, who accounted for more than 60% of the total athletes. About 32% of the athletes travelled to Langkawi with at least two or three other people and stayed for five to ten days. Approximately 93% of the athletes said they stayed in a hotel.

For instance, during the Ironman 2014 edition, international athletes spent an average of US$2321 during their stay in Langkawi. Accommodation, tour activities and shopping were the top three expenses. For meals and car rentals, the bulk of the athletes paid at least US$25 a day. Overall, the event created a direct effect of US$0.4 million from participation fees and a US$2.3 million indirect impact on Langkawi's economy. Using this as an example, it comes as no surprise that sporting events are Langkawi's main tourist product, with at least one event held every month.

Conclusion

In a nutshell, sporting events enhance the destination marketing strategy and have a long-term effect on the tourism industry. Because of their direct and indirect economic impacts, sporting events are a major tourism segment for emerging economies. This chapter has provided some case studies on sporting events and destinations in Asia by examining the marketing and economic effects of sporting events on the nation's economy. The findings could help the government allocate funds to promote sporting activities and boost tourist destination services by emphasizing target marketing campaigns.

References

Andersen, S. S., Hanstad, D. V., & Plejdrup-Skillestad, K. (2015). The role of test events in major sporting events. *Event Management, 19*(2), 261–273.

Asia Outlook. (2019). RWC 2019: *Japan's $1.8 billion bonanza.* https://www.asiaoutlookmag.com/industry-insights/article/771-rwc-2019-japans-18-billion-bonanza

Association of South East Asia Nations (ASEAN). (2017). *Tourism marketing strategy 2017–2020.* https://asean.org/wpcontent/uploads/2012/05/ASEAN_Tourism_Marketing_Strategy_2017-2020.pdf

China Business Review. (2008). *The 2008 Olympics' impact on China.* https://www.chinabusinessreview.com/the-2008-olympics-impact-on-china/

COMCEC. (2015). Effective tourism marketing strategies: ICT-based solutions for the OIC member countries. In *Proceedings of the 6th Meeting of the COMCEC Tourism Working Group,* COMCEC.

Deloitte. (2020). *Sponsoring sports in today's digital age.* https://www2.deloitte.com/content/dam/Deloitte/lu/Documents/sports-consulting/lu-sponsoring-sports-today-digital-age.pdf

Foris, D., Florescu, A., Foris, T., & Barabas, S. (2020). Improving the management of tourist destinations: A new approach to strategic management at the DMO level by integrating lean techniques. *Sustainability, 12*(23), 10201. https://doi.org/10.3390/su122310201

Getz, D. (1989). Special events: Defining the product. *Tourism Management, 10*(2), 125–137.

Getz, D. (2008). Event tourism: Definition, evolution, and research. *Tourism Management, 29,* 403–428.

Getz, D. (2012). Event studies: Discourses and future directions. *Event Management, 16*(2), 171–187.

Hamid, Z. A., Khairuddin, N. I. M., Mahdzar, M., & Anuar, N. A. M. (2017). Interactive tourism websites attributes and tourist decision to visit Malaysia. *Journal of Tourism, Hospitality and Culinary Arts, 9*(2), 47–58.

Heras-Pedrosa, C. D. L., Millan-Celis, E., Iglesias-Sánchez, P. P., & Jambrino-Maldonado, C. (2020). Importance of social media in the image formation of tourist destinations from the stakeholders' perspective. *Sustainability, 12*(10), 4092. https://doi.org/10.3390/su12104092

Hover, P., Dijk, B., Breedveld, K., van Eekeren, F. J. A., & Slender, H. (2016). *Creating social impact with sport events*. WJH Mulier Instituut.

Hussin, N. Z. I., & Buchmann, A. (2019). Understanding tourism development policies in Malaysia. *Journal of Policy Research in Tourism, Leisure and Events, 11*(2), 333–353.

International Olympic Committee. (2018). *PyeongChang Olympic winter games create legacy of environmental and social good.* https://olympics.com/ioc/news/pyeongchang-olympic-winter-games-create-legacy-of-environmental-and-social-good

ISC. (2020). *Case study Malaysia Truly Asia.* https://www.iscsea.com/case-study-malaysia-truly-asia

Kent County Council. (2016) *Sports events checklist. Sports sheets.* https://www.kentsport.org/wp-content/uploads/2016/06/sports_sheet_sports_events_checklist.pdf

Khor, P. H., Lim, K. C., & Radzuwan, R. (2019). The sports event's quality towards attendance of tourists at sports event hosted at Northern Zone of Malaysia/Khor Poy Hua, Lim Khong Chiu and Radzliyana Radzuwan. *Journal of Academia, 7*(1), 126–132.

Kim, E. J., & Kang, Y. (2020). Spillover effects of mega-events: The influences of residence, transportation mode, and staying period on attraction networks during Olympic games. *Sustainability, 12*(3), 1206. https://doi.org/10.3390/su12031206

Knott, B., & Hemmonsbey, J. (2015). The strategic value of sport for an African city brand. *African Journal for Physical, Health Education, Recreation and Dance, 21*(1), 191–204.

Kruja, D. (2018). Destination marketing research. In D. Gursoy & C. G. Chi (Eds.), *The Routledge handbook of destination marketing* (pp. 35–48). Routledge.

Lintumäki, P., Winner, H., Scheiber, S., Mederle, A., & Schnitzer, M. (2020). The economic impact of participant sports events: A case study for the winter world masters games 2020 in Tyrol, Austria. *Economies, 8*(4), 94. https://doi.org/10.3390/economies8040094

Malaysia Convention & Exhibition Bureau (MyCEB). (2015). *Annual report.* https://www.myceb.com.my/clients/Malaysia_Convention_and_Exhibition_Bureau_75E277C6-C63D-4DC3-89AC-F6B30D21417D/contentms/img/pdf/Report/2015_myceb_annual_report.pdf

Malaysia Convention & Exhibition Bureau (MyCEB). (2018). *Annual report.* https://www.myceb.com.my/clients/Malaysia_Convention_and_Exhibition_Bureau_75E277C6-C63D-4DC3-89AC-F6B30D21417D/contentms/img/pdf/Report/2018_myceb_annual_report.pdf

Malaysia Convention & Exhibition Bureau (MyCEB). (2020). *Malaysia's business events roadmap: Charting Malaysia's journey to 2020 and beyond.* https://www.myceb.com.my/clients/Malaysia_Convention_and_Exhibition_Bureau

Mapjabil, J., Marzuki, M., Kumalah, M. J., Tangavello, L., & Abidin, M. K. Z. (2017). Sport as a tourism attraction in Malaysia: Potential and prospects. *Geografia-Malaysian Journal of Society and Space, 11*(12), 23–31.

Mihajlović, I. (2017). The importance of local events for positioning of tourist destination. *European Journal of Social Science Education and Research, 4*(4), 228–239.

Müller, M. (2015). What makes an event a mega-event? Definitions and sizes. *Leisure Studies, 34*(6), 627–642.

Nguyen, T. T. N. (2017). The role of the special event in destination marketing. In *Proceedings of International Business Management Conference.* Association for Computing Machinery.

Nufer, G. (2016). Sports events and social media marketing. *International Journal of Economics, Commerce and Management, 4*(3), 16–28.

Oklobdžija, S. (2015). The role of events in tourism development. *Bizinfo (Blace), 6*(2), 83–97.

Pitts, B., & Stotlar, D. (2013). *Fundamentals of sport marketing* (4th ed.). FiT Publishing.

Sato, T., Fukumoto, S., Ishii, M., Hiraishi, I., & Matsumoto, Y. (2018). Digital marketing that leverages sporting event ticketing data. *Fujitsu Scientific & Technical Journal, 54*(4), 32–37.

Siyabulela, N. (2016). Using sport tourism events as a catalyst for tourism development in the Eastern Cape Province, South Africa. *African Journal of Hospitality, Tourism and Leisure, 5*(3), 1–12.

Sotiriadis, M. (2021). Tourism destination marketing: Academic knowledge. *Encyclopaedia, 1*(1), 42–56. https://doi.org/10.3390/encyclopedia1010007

Statista. (2020). *Number of inbound visitors to South Korea from 2000 to 2020.* https://www.statista.com/statistics/709116/south-korea-inbound-visitors/

The Organization for Economic Co-operation and Development (OECD). (2018). *OECD tourism trend and policies 2018.* https://www.oecd-ilibrary.org/docserver/tour-2018-6-en.pdf

Tourism Malaysia. (2014). *Tourism Malaysia launches integrated digital marketing programme.* https://www.tourism.gov.my/pdf/uploads/

Tourism Malaysia. (2016). *Citrawarna@Kuala Lumpur 2016 unleash the moment, unfold the stories.* https://www.tourism.gov.my/pdf/uploads/

Wafi, A. A., Lim, K. C., & Ibrahim, J. A. (2018). Small-scale event sports tourism as visitors' attractions: The case of northern states in Malaysia. *Ottoman: Journal of Tourism and Management Research, 3*(2), 261–273.

Wang, Y., & Jin, X. (2019). Event-based destination marketing: The role of mega-events. *Event Management, 23*(1), 109–118.

4

Competitiveness of Emerging Economies as Tourist Destinations

Emmanuel Gamor and Ishmael Mensah

Introduction

The competitiveness of a destination is "its ability to increase tourism expenditure, to increasingly attract visitors while providing them with satisfying, memorable experiences, and to do so in a profitable way, while enhancing the wellbeing of destination residents and preserving the natural capital of the destination for future generations" (Ritchie & Crouch, 2003, p. 2). The competitiveness of emerging economies in the global tourism market has become even more imperative, as many countries have intensified efforts at gaining a share of the market. However, the competitiveness of tourist destinations depends not only on the ability to

E. Gamor (✉)
Hong Kong Polytechnic University, Hung Hom, Hong Kong

Koforidua Technical University, Koforidua, Ghana
e-mail: egamor@ktu.edu.gh

I. Mensah
University of Cape Coast, Cape Coast, Ghana
e-mail: ikmensah@ucc.edu.gh

I. Mensah et al. (eds.), *Marketing Tourist Destinations in Emerging Economies*, Palgrave Studies of Marketing in Emerging Economies,
https://doi.org/10.1007/978-3-030-83711-2_4

attract and satisfy tourists but also on how that results in the improved well-being of host communities. Emerging economies have comparative advantages in their natural and cultural resource endowments, but that alone does not guarantee competitiveness. What is important is how the destination can use these resources effectively by creating and managing other facilities, services, and infrastructure as well as providing the right policy framework for tourism development. Despite their huge tourism potentials, emerging economies have not yet managed to fully exploit their tourism potential (Sanches-Pereira et al., 2017) due to some challenges such as poor infrastructure, ineffective marketing, poor linkages between tourism and other sectors of the local economy, weak institutional and technical structures, and lack of qualified personnel. Thus, the development of attractive destination resources, branding, formulation of regulatory policies, planning, and adoption of technology are critical success factors for emerging economies seeking to remain competitive in the global tourism market. This chapter delves into issues relating to the competitiveness of emerging destinations and the implications for marketing. Specifically, the chapter looks at the concept of destination competitiveness, comparative and competitive advantage of emerging destinations, the competitiveness of tourist destinations in emerging economies, factors influencing the competitiveness of emerging destinations, and critical success factors in destination marketing in emerging economies.

Destination Competitiveness

The competitiveness of a tourist destination is considered a crucial factor for its success (Buhalis, 2000; Dwyer & Kim, 2003). It is the competitiveness of a destination that ultimately determines its attractiveness to both tourists and investors. Destination competitiveness has, therefore, become a very important issue, which constantly engages destination management organisations (DMOs) and governments. The need for destinations to be competitive has become even more imperative due to the heightened competition in the global tourism marketplace. The increasing competition among tourist destinations has been attributed to the emergence of many new destinations, including emerging economies onto the market while the main generating markets remain unchanged (Acemoglu & Robinson, 2012; Candela & Figini, 2012). The situation has been worsened by recent crises, including economic recession, natural disasters, and pandemics like COVID-19. This has placed tourism destinations under extreme pressure to differentiate themselves from competitors, become unique and recognisable, and remain competitive (Krešić & Prebežac, 2011).

Competitiveness has been described by Dwyer and Kim (2003) as a multifaceted concept. In view of this, it could be examined from different perspectives, such as products, companies, sectors of the economy, or national economies, in both the short and long run (Omerzel, 2006). Competitiveness is the degree to which a country or an organisation produces goods and services that meet the standards of the market (especially the international markets) while at the same time maintaining and enhancing the benefits in terms of real income to its citizens, managers, or shareholders (Krugman, 1994; Waheeduzzaman, 2011). However, in the context of the tourist destination, competitiveness has been viewed mainly from the supply side to include the ability of the tourist destination to attract and satisfy tourists by delivering better products and services compared to other competing destinations (Aqueveque & Bianchi, 2017; Crouch & Ritchie, 1999; Dwyer & Kim, 2003), and maintain and improve its market share and competitive position (Alseiari et al., 2019; d'Hauteserre, 2000). In this vein, Dwyer and Kim (2003) posit that a competitive destination has to be more successful than competitors in various aspects of the overall tourist experience. Thus, the competitive advantage of a destination lies in the strength or ability of the destination to provide a quality experience to visitors (Ritchie & Crouch, 2003).

A competitive destination has also been described as a destination that is price competitive (Dwyer et al., 2000). Dwyer et al. (2000) described destination competitiveness as encompassing price differentials coupled with exchange rate movements, productivity, and qualitative factors which affect the attractiveness of a destination. However, because destinations have different stakeholders including host communities, competitiveness should not only be viewed from the perspective of tourists and tourism businesses alone. It must include the broader socio-economic and environmental impacts of tourism at the destination. In this vein, Ritchie and Crouch (2003, p. 2) provide a more all-encompassing definition of the competitiveness of a destination as:

> its ability to increase tourism expenditure, to increasingly attract visitors while providing them with satisfying, memorable experiences, and to do so in a profitable way, while enhancing the wellbeing of destination residents and preserving the natural capital of the destination for future generations.

Thus, the competitiveness of a tourist destination does not only depend on its ability to attract tourists over the long term but also depend on its ability to protect its tourism resources and enhance the quality of life of its citizens (Rey-Maquieira & Ramos, 2006). Destination competitiveness must reflect in the economic prosperity and quality of life of the residents of a country (Buhalis, 2000; Crouch & Ritchie, 1999; Dwyer & Forsyth, 2003; Dwyer & Kim, 2003). Dwyer and Kim (2003, p. 372) emphasise that the ultimate goal of a destination's competitiveness is "to maintain and increase the real income of its citizens, usually reflected in the standard of living of the country". This is especially important for emerging economies, where the creation of sustained and quality jobs, enhanced local economic growth, improved social well-being, and improved living standards are usually overlooked in favour of tourism investments and revenue.

Indicators of Destination Competitiveness

Models of Destination Competitiveness

The basis for competitiveness is Porter's (1990) diamond model of competitiveness. Porter provided four broad attributes of competitiveness of an industry or a company, namely, factor conditions, referring to the supply of skilled labour or infrastructure; demand conditions; related and supporting industries; and firm strategy, structure, and rivalry. Based on Porter's (1990) model, Crouch and Ritchie (1999) developed a conceptual model of destination competitiveness. In their model, they identified four main determinants of destination competitiveness, namely the core resources and attractions of a destination including climate, special events, and tourism facilities; supporting factors and resources such as infrastructure; destination management, including the activities of destination management organisations (DMOs); and broader situational conditions such as safety, government policies, and environmental circumstances which could negatively affect competitiveness (Crouch & Ritchie, 1999; Ritchie & Crouch, 2003). Crouch and Ritchie (1999) further developed a comprehensive list of indicators that combine subjective consumer and objective industry indicators for each of the components, which culminated in a composite destination competitiveness index.

Crouch and Ritchie's (Crouch & Ritchie, 1999; Ritchie & Crouch, 2003) competitiveness model provided a good basis for the development of the integrated model of destination competitiveness by Dwyer and Kim (2003). Dwyer and Kim pointed out that a major weakness of the Ritchie and Crouch's conceptual model was its neglect of demand-side determinants of competitiveness and its focus on supply-side elements. They, therefore, proposed a more holistic set of indicators of destination competitiveness which were categorised into five subgroups, namely, endowed resources, such as mountains, landscapes, and cuisine; supporting factors, including infrastructure, service quality, and accessibility; destination management, such as marketing strategy and safety regulations; situational conditions such as crimes, and global economic conditions; and demand factors, including awareness, perception, and tourist preferences.

In practical terms, the competitiveness of tourist destinations has been conceptualised variously. Different measures have been used to gauge destination competitiveness because of the varied interpretations placed on the concept. According to Kozak and Rimmington (1999), in general terms, there are two distinct approaches to measuring destination competitiveness: the use of "hard data" and "soft data". Hard data refers to objectively verifiable data such as tourist arrivals, tourism receipts, market share, tourist expenditure, and employment, while soft data refers to subjective data such as quality of tourism services, the authenticity of culture, and tourist satisfaction. There appears to be no agreement as to the number of indicators that should be used to measure the competitiveness of tourist destinations. The competitiveness of tourist destinations is derived from a combination of several factors including economic, political, psychological, social/cultural, environmental, and increasingly more important, technological factors (Mulec & Wise, 2013). Bernini (2009) provides four main factors, namely, demand conditions, such as the quality, level, and nature of tourist interest; local factor conditions, such as a destination's natural resources and location; tourism-related and supporting industries; and government policies for supporting communities with information and infrastructure. Previous studies on destination competitiveness have employed a plethora of factors, including destination resources and destination management, as well as demand, socio-economic, political, and environmental factors. According to Kozak et al. (2009), tourists choose destinations by comparing these factors among various destinations.

Perhaps the most comprehensive set of indicators for gauging the competitiveness of a destination is the travel and tourism competitiveness index (TTCI), which was developed by the World Economic Forum (WEF) in 2004. It consists of a set of 14 pillars categorised into three subgroups, namely travel and tourism regulatory framework; travel and tourism business environment and infrastructure; and travel and tourism human, cultural and natural resources. The travel and tourism regulatory framework subgroup has five pillars, namely, A1 (Policy rules and regulations), A2 (Environmental sustainability), A3 (Safety and security), A4 (Health and hygiene), and A5 (Prioritisation of travel and tourism). The travel and tourism business environment and infrastructure subgroup also has five pillars, namely B1 (Air transport infrastructure), B2 (Ground transport infrastructure), B3 (Tourism infrastructure), B4 (ICT infrastructure), and B5 (Price competitiveness in the T&T industry). The third subgroup (travel and tourism human, cultural, and natural resources) has four pillars, namely, C1 (Human resources), C2 (Affinity for travel and tourism), C3 (Natural resources), and C4 (Cultural resources). Based on the TTCI, the WEF ranks 124 countries (WEF, 2015a).

Comparative and Competitive Advantage of Emerging Destinations

The competitiveness of tourist destinations borders on not just the availability of resources and attractions, but also how the resources are harnessed and utilised. Thus, destinations need both comparative and competitive advantages. Comparative advantages relate to the resources that are available to a destination (resource endowments), while competitive advantages refer to the destination's ability to utilise the available resources (resource deployment) (Crouch & Ritchie, 1999; Porter, 1990). The comparative advantage of a tourist destination refers to its endowed or inherited resources, including beaches, climate, mountains, wildlife, heritage sites, and folklore, while competitive advantage refers to what the destination has created to ensure the effective use of the resources such as hotels, transport networks, events, the quality of management, skills, and competencies of workers, and government policy. Comparative advantage alone does not guarantee a destination's competitiveness. A destination with considerable resource endowments will require the requisite superstructure and managerial know-how to take full advantage of those resources. According to Crouch and Ritchie (1999), this means that a country that is endowed with a wealth of resources might be less competitive than a destination that lacks resources but uses the ones it does have more effectively.

In spite of this, most emerging economies tend to capitalise on their comparative advantages to remain competitive in the global tourism marketplace. In a study on the competitiveness of the South Banat District of Serbia, core resources and attractors had the highest ratings out of four competitiveness determinants, with gastronomy, multicultural ambience, the richness of cultural and historical heritage, and climate having the highest ratings (Kovačević et al., 2018). Also, in a study on the competitiveness of the Vojvodina region of Serbia, Mulec and Wise (2013) found that though the region possesses much in terms of natural and cultural resources, they concluded that more investments, marketing, strategic management, and planning are needed to make the destination more attractive to international visitors and to improve the region's competitiveness. In a comparative study of the competitiveness of the tourism industry in Serbia and Slovenia by Armenski et al. (2011), the results showed that both destinations were considered to be more competitive in natural, cultural, and created resources but were less competitive in destination management with unfavourable demand conditions. The WEF (2012) has indicated that though ASEAN as a whole has a price competitive advantage compared with the rest of Asia and other parts of the world, the region as a whole, except Singapore, has some flaws including inadequate infrastructure, unsatisfactory sanitation, and environmental degradation. Azarya (2004) asserts that international tourists' desire for experiencing different cultures, landscapes, and wilderness whilst remaining in the familiar comforts and security of their home environment worsens the competitive challenge in developing countries that lack adequate supporting factors. This means that for emerging economies to remain competitive, they must develop the requisite supporting superstructure, formulate relevant policies, and manage their natural and cultural resources more effectively. Singapore provides a classic case of how an emerging economy could become competitive by capitalising on its competitive advantage. Though Singapore is not endowed with abundant natural resources, it ranks among the top ten countries regionally, in the Asia Pacific region, according to the Travel and Tourism Competitiveness Index (World Economic Forum, 2015a). Singapore's impressive performance is attributed to its planning excellence in a wide range of tourism-related projects including the modernisation of its economic infrastructure, economic liberalisation and international openness, innovative management in developing natural and cultural resources, and high-quality human resources development (Ganguli & Ebrahim, 2017).

Competitiveness of Emerging Tourist Destinations

It has been acknowledged that the destination competitiveness of emerging economies is imperative for the growth of their economies (Aqueveque & Bianchi, 2017; Echtner & Ritchie, 2003). Through job creation and revenue and income generation, tourism is seen as a vehicle for the attainment of socio-economic development in emerging economies. As a result, many destinations from emerging regions of the world are attempting to develop and exploit their tourism potential to attract and cater for visitors from both domestic and international markets (Aqueveque & Bianchi, 2017). However, for emerging economies to derive the desired benefits from tourism development, they must be competitive in the global tourism market. Evidence from the TTCI, however, indicates that the top ranks of the index are dominated by advanced economies, though emerging economies have shown some promise in recent years (Table 4.1).

Table 4.1 Ranking of countries based on the travel and tourism competitiveness index (2015–2019)

	Rank			
Country	2015	2017	2019	Increase/decrease in rank (2017–2019)
Spain	1	1	1	–
France	2	2	2	–
Germany	3	3	3	–
USA	4	6	5	Increase
UK	5	5	6	Decrease
Switzerland	6	10	10	–
Australia	7	7	7	–
Italy	8	8	8	–
Japan	9	4	4	–
Canada	10	9	9	–
Singapore	11	13	17	Decrease
Austria	12	12	11	Increase
Hong Kong SAR	13	11	14	Decrease
Netherlands	14	17	15	Increase
Portugal	15	14	12	Increase
New Zealand	16	16	18	Decrease
China	17	15	13	Increase
Iceland	18	25	31	Decrease
Ireland	19	23	26	Decrease
Norway	20	18	20	Decrease
Belgium	21	21	24	Decrease
Finland	22	33	28	Increase
Sweden	23	20	22	Decrease
UAE	24	29	33	Decrease
Malaysia	25	26	29	Decrease
Luxemburg	26	28	23	Increase

(continued)

Table 4.1 (continued)

Country	Rank			Increase/decrease in rank (2017–2019)
	2015	2017	2019	
Denmark	27	31	21	Increase
Brazil	28	27	32	Decrease
Korea Rep.	29	19	16	Increase
Mexico	30	22	19	Increase
Greece	31	24	25	Decrease
Taiwan, China	32	30	37	Decrease
Croatia	33	32	27	Increase
Panama	34	35	47	Decrease
Thailand	35	34	31	Increase
Cyprus	36	52	44	Increase
Czech Republic	37	39	38	Increase
Estonia	38	37	46	Decrease
Slovenia	39	41	36	Increase
Malta	40	36	35	Increase
Hungary	41	49	48	Increase
Costa Rica	42	38	41	Decrease
Qatar	43	47	51	Decrease
Turkey	44	44	43	Increase
Russian Federation	45	43	39	Increase
Barbados	46	58	a	–
Poland	47	46	42	Increase
South Africa	48	53	61	Decrease
Bulgaria	49	45	45	–
Indonesia	50	42	40	Increase

[a]Not covered due to insufficient data
Sources: WEF, TTCI (2015b, 2017, 2019a)

Since 2015, Singapore has been the most competitive emerging economy based on the TTCI, as it was ranked 11, 13, and 17 in 2015, 2017, and 2019, respectively. The best performing emerging economies are mainly found in the Asia Pacific region, where countries like China, Hong Kong SAR, Singapore, and Malaysia ranked relatively high. The only African country ranked among the top 50 in 2015 was South Africa, but it dropped to 61 in 2019. Experts have predicted that inbound tourism in the Asia Pacific region will grow at a faster pace than in the advanced economies in the coming decades. The United Nations World Tourism Organization (UNWTO) predicted that arrivals will have increased by

331 million to 535 million in 2030 (UNWTO, 2011). However, the COVID-19 pandemic which virtually grounded tourism to a halt in 2020 will seriously hamper the realisation of this projection. Emerging economies will need more time to recover from the devastating effect of COVID-19. Qiu et al. (2021) posit a recovery of tourism in the Asia Pacific in 2020, ranging from 10% to 70%, that will serve as the basis for full recovery in the subsequent years, as conditions improve gradually. Emerging economies accounted for only US$ 486 billion out of the US$ 1260 billion international tourism receipts in 2016 (Goffi et al., 2019). In spite of these, emerging economies have not yet managed to fully exploit their tourism potential (Sanches-Pereira et al., 2017). Several factors hinder the competitiveness of emerging tourist destinations. These include poor infrastructure, lack of new product development and management, ineffective marketing, poor linkages between tourism and other sectors of the local economy, poor institutional and technical capabilities, and shortage of skilled and qualified personnel (Mitchell & Faal, 2007). Tourism has been developed in most emerging economies without an integrated plan and attention to the environmental impacts and impacts on local communities (Goffi et al., 2019). There is also the problem of lack of coordination and stakeholder collaboration, which makes coordinating the efforts of a variety of independent stakeholders towards a common goal a difficult task (Laws et al., 2011). However, for the sustainable competitiveness of tourist destinations to be realised, major stakeholders at the destination including national tourism organisations, DMOs, and local tourism business operators must work together. This is due to the composite nature of the tourist product and the multidimensional structure of the industry.

Dwyer and Kim (2003, p. 399) opine that "there is no single or unique set of competitiveness indicators that apply to all destinations at all times". Emerging economies are in a unique position; therefore, the indicators for gauging the competitiveness of more established destinations may not be suitable for such economies. The major tourism destination competitiveness models have been developed purposely for the advanced countries and therefore did not take into consideration the peculiar needs of emerging economies (Azzopardi & Nash, 2016). The TTCI, for instance, may be useful to large countries that are capable of targeting diverse market segments because their size and resources afford them the capability to pursue such strategies, but smaller destinations tend to pursue a more selective targeted marketing strategy and are only interested in information that will enhance their competitiveness in certain specific market segments (Croes, 2010; Crouch, 2008).

Malaysia, for instance, has specifically targeted Middle Eastern tourists to the country through its "Look East" policy and its religious connections to those countries. The former Prime Minister Tun Dr Mahathir capitalised on the vacuum created in the aftermath of the 11 September 2001 attack on the USA, as Middle Eastern tourists felt uncomfortable travelling to Western countries due to the mistreatment they received. As a result, Malaysia has become one of the favourite alternative vacation destinations for Arabs (Ibrahim et al., 2009). Another country that has capitalised on its comparative advantage to target a specific market segment is Ghana. Ghana has capitalised on its comparative advantage in resources and attractions relating to the transatlantic slave trade (TAST) to target the African diaspora. Among the TAST resources are slave castles which are world heritage monuments, other landmarks and relics, special events like PANAFEST, and the historical connections with the African Diaspora through the Pan African movement, which was championed by the first president of Ghana, Dr Kwame Nkrumah.

The reality is that countries have different challenges and priorities as they evolve from resource-based to knowledge-based economies. Thus, the principal factors that contribute to global competitiveness, for economies at different levels of development differ (Porter et al., 2001). Most emerging economies are resource-based and, therefore, their main sources of competitiveness are their natural resource endowments. As a result, the indicators of competitiveness of more established destinations may not sit well with the peculiarities of emerging economies. In a study on the destination competitiveness of Chile, by Aqueveque and Bianchi (2017), awareness and promotion of the destination in foreign markets were considered very relevant for destinations in emerging economies. They, therefore, recommended that future analyses of destination competitiveness in developing countries must include this aspect of competitiveness.

COVID-19 has had a significant impact on the global economic, political, and social systems. Since its declaration as a worldwide pandemic by the WHO in March 2020, it has negatively affected the tourism sector, as it gradually grounded international travel to a halt. Emerging destinations suffered significantly from the pandemic. For instance, a major activity that attracts tourists to South Africa, the 51st edition of The Two Oceans Marathon", was cancelled (Swart & Maralack, 2020). The desire to travel dropped by 50% in Asia, while in Europe and America, it dropped by some 30% (Gallego & Font, 2020). The COVID-19 pandemic has caused fear globally, impacting destination competitiveness. Destinations that are able to reach herd immunity through vaccination, together with the health and safety protocols, will be more competitive than other destinations. Emerging destinations in this era are more likely to become less competitive, considering the COVID-19 situation. In view of the vaccination cover, the USA has 46.23% of the population vaccinated against COVID-19, Spain has 31.13%, and Canada has 42.29% as of 13 May 2021. However, emerging economies have lower coverage of vaccination. For example, Brazil has 16.04%, Ghana, 2.74%, India, 10.08%, Hong Kong, 12%, and South Africa, 0.77% (U.S. Department of Health & Human Services, 2021). With this, the emerging economies are less likely to be competitive with respect to tourist safety and other health protocols related to COVID-19.

Factors Influencing the Competitiveness of Emerging Destinations

The continuous attempt to improve destination competitiveness due to the changing nature of destinations rests upon varied reasons and factors (Cronjé & du Plessis, 2020). The multiple goals of competitiveness make it difficult to conceptualise (Goffi et al., 2019). As such, various factors are usually ascribed to the realisation of the competitiveness of a tourist destination as shown in Fig. 4.1. Emerging destinations are heterogeneous and target different travel market segments, and factors relating to their competitiveness are also varied, yet little attention has been given to the elements that determine the competitiveness of emerging

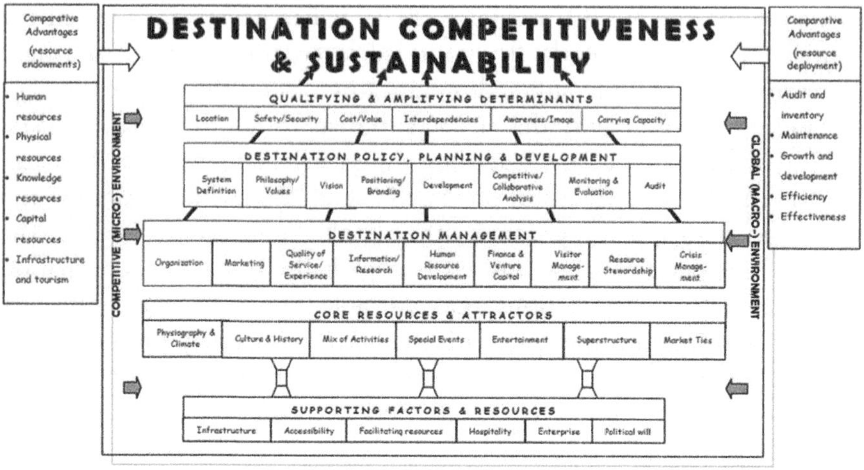

Fig. 4.1 Destination competitiveness model. (Source: Ritchie and Crouch (2003))

destinations. As a result, most of the determinants of the competitiveness of tourist destinations in the literature are largely based on studies in the developed economies, although emerging economies are gradually dominating the international tourism marketplace (Goffi et al., 2019).

Competitiveness arises from the position that a destination occupies in the global travel market due to its performance and the potential to attain a higher position with all the strategies employed (Goffi et al., 2019; Kovačević et al., 2018; Papp & Raffay, 2011; Ritchie & Crouch, 2003). The elements of competitiveness are categorised into major determinants including environmental uniqueness and sustainability (Goffi et al., 2019; Hassan, 2000), institutional linkages and prices (Heath, 2002), demand conditions (Dwyer & Kim, 2003), and performance (World Economic Forum, 2019). However, performance models such as the TTCI have often been criticised by researchers as being vague and do not address the practical issues relating to competitiveness (Kovačević et al., 2018) especially in emerging economies.

To thrive in the ever-competitive environment, emerging economies need to understand the approaches and factors used in measuring their level of competitiveness and develop appropriate strategies for improving their positions. The factors influencing destination competitiveness in emerging economies relate to both the comparative and competitive advantages (Magombo et al., 2017; Papp & Raffay, 2011) and go beyond the attributes of the destination. These factors help destinations to identify what contributes to their strongest and weakest position on the market. The following paragraphs present major factors acknowledged to influence destination competitiveness in emerging economies.

Core Resources. Resource endowment is one of the main factors easily identified and used as a measure of the competitiveness of destinations. They are made up of natural and created resources (Ritchie & Crouch, 2003), which serve as the main pull factor that attracts tourists to a destination. Unique attributes of a destination usually come from inherited resources (Dwyer & Kim, 2003) that afford the destination a competitive edge over other destinations. The resource endowment of a destination refers to the essential tourism resources or attractors such as the flora, fauna, climate, and landscape. Michael et al. (2019) argue that because essential resources of a destination have a significant influence on the destination's competitiveness, core resources cannot be overemphasised. Goffi et al. (2019) are of the view that the main tourism resources and the services provided to tourists offer value and form the main motivation behind travel for tourism activities. Core resources include the physiography and climate, special events, entertainment, and activities, as well as the culture and history of a destination (Ritchie & Crouch, 2003). Therefore, the uniqueness of core resources is used as differentiating and competitive factor relative to other destinations (Salinas Fernández et al., 2020).

Supporting Factors. The core resources function well with the help of supporting factors. These factors make the travel to and stay in a destination comfortable and memorable, and provide a foundation for successful industry development (Ritchie & Crouch, 2003). Enright and Newton (2004) claim that the support factors include destination infrastructure, educational institutions, and accessibility. Hospitality, quality of service, accessibility, and market ties also play an important role in the competitiveness of a destination as support factors (Crouch, 2011; Dwyer & Kim, 2003; Heath, 2002). Good roads, airports, bus systems, and telecommunications, among others, propel the appeal and increase the chance of a destination being chosen over its competitors. A destination that is perceived as friendly and warm makes tourists feel welcome. Hospitality, therefore, is a very important social factor that determines the competitiveness of a destination. Similarly, the linkages that destinations build with source markets such as diaspora relations between Ghana and the USA, and complementary linkages with other destinations such as the reliance of Macao SAR on Hong Kong SAR, contribute to the competitiveness of a destination. In general, the supporting factors are deemed very important (Salinas Fernández et al., 2020) in determining competitiveness through establishing a strong and successful industry.

Situational Conditions. Building, maintaining, and amplifying a solid image and capability of satisfying tourists is important to destinations. Hence, it is important to consider factors that can negatively change the effect of core attractors and supporting factors towards destination competitiveness. Also known as amplifying and qualifying factors (Ritchie & Crouch, 2003), the situational conditions include factors of the macro environment and the operational environment such as location, operational capabilities of tourism businesses, safety and security, technology readiness, and price competitiveness (Kovačević et al., 2018). Situational factors can either improve or limit the competitiveness of a destination (Enright & Newton, 2004; Salinas Fernández et al., 2020). Currency exchange rate, the legal regulatory environment, and level of visitor safety and security are some of the factors that can negatively influence the competitive position of a destination if not well managed.

Destination Management. To successfully maintain a competitive edge over other destinations, DMOs need to carefully and systematically examine the services and resources that place their destinations at a unique and advantageous position over other destinations (Dwyer & Kim, 2003; Hassan, 2000; Heath, 2002). As Khalifa (2020) acknowledges, diligent management of destinations supports their competitiveness. Destination management is a multifaceted activity that involves destination planning, destination marketing, destination management organisation, managing the environment, human resource development, crisis management, visitor management, and resource stewardship (Crouch, 2011; Dwyer & Kim, 2003; Ritchie & Crouch, 2003). These activities, according to Crouch (2011), are carried out collectively by organisations and individuals in the interest of promoting the brand, image, and enhancing the position of the destination through well-coordinated planning and organisation of human resource and sustainable development. In this case, it is important to guard against devaluation and deterioration of the resources to ensure prolonged benefits to destinations. To do so, training and development and commitment to the environment is an important issue to consider.

Creating Competitive Emerging Destinations: Critical Success Factors

Many governments in developing economies are focusing on tourism as a source of foreign exchange earnings, which is vital for improving their economies (Kankhuni, 2020; Khalifa, 2020). Thus, understanding the critical factors of destination competitiveness among emerging economies is important for tourism development. For emerging economies to fully exploit their tourism potentials and remain competitive, their destination marketing strategies should be informed by some critical success factors including attractive destination resources, branding, regulatory policy, planning and development, and technology adaptation.

Attractive Destination Resources. The competitiveness of emerging destinations depends on the uniqueness and attractiveness of tourism resources which are the main reasons for visiting such destinations (Ferreira & Perks, 2020; Kovačević et al., 2018; Michael et al., 2019; Ritchie & Crouch, 2003). These include the attractions, activities, and special events. These constitute *raison d'etre* for choosing one destination over another (Kovačević et al., 2018; Ritchie & Crouch, 2003). Destination attributes are central to the competitiveness of destinations because they form the pivot around which tourism activities revolve and the basis of the experience acquired by the tourist. Thus, without attractions, there will be no need for people to travel to certain destinations. Also, the environmental and climatic conditions of an area enable certain activities and events and, thus, tend to attract tourists to such destinations (Ferreira & Perks, 2020). Notably, the climate of the tropics and the Mediterranean region attracts tourists from temperate zones (mostly developed economies) at certain times of the year. To this end, the availability of unique resources contributes to the competitiveness of emerging economies.

Emerging economies are usually endowed with unique natural resources and activities that attract visitors. These include nature-based or water-based activities, festivals, lakes, mountains, idyllic climate, flora, and wildlife. To gain a competitive advantage, emerging economies should develop these core resources as their main attractions. Indeed, in most emerging economies such as Malawi, Serbia, Ghana, and UAE, tourism is driven by attractive resources, activities, and events (Kankhuni, 2020; Kovačević et al., 2018). In the UAE, for instance, the tourism resources of the destination, including historical buildings in heritage villages and cities, rocky mountains, and religious buildings like the Sharia Mosque, are viewed as one of the most important determinants of its competitiveness (Michael et al., 2019). Similarly, Kovačević et al. (2018) argue that gastronomy, multicultural ambience, and the richness of the Serbian culture and historical heritage determine Serbia's competitiveness. Also, in the Autonomous regions of Madeira, activities that are interrelated with artisan culture are extremely important to the competitiveness of the region (Teixeira & Ferreira, 2019). More so, special events such as festivals, conventions, and exhibitions rank among the top factors accounting for the competitiveness of emerging economies (Cronjé & du Plessis, 2020; Goffi et al., 2019).

Tourism Services and Infrastructure. The second critical success factor identified is the quality of infrastructure and tourism services. The transport network in a destination contributes to its competitiveness in the market. This factor depends on the political will and the strategic planning of the government of the destination. As noted by Michael et al. (2019), the availability and accessibility of hotels, restaurants, retail services, and the quality of road and railroad networks, sanitation, quality of electricity supply, quality of health care, improved internet access, and water supply are pertinent to the gaining advantage over competitors in the market. Recent literature has shown that supporting services, together with general and tourism infrastructure such as the availability of airports, the number of air, water, and road connections that facilitate international arrivals, and tourism superstructure, play a very crucial role in enhancing the attractiveness of an emerging economy as a tourist destination (Cronjé & du Plessis, 2020; Michael et al., 2019).

To Cronjé and du Plessis (2020), infrastructure and its development have been identified by multiple scholars of both developed and emerging economies as crucial to the success of destination competitiveness. Destinations with good and accessible infrastructure and other services have gained an advantage as very competitive destinations. Kim, Arcodia, and Kim (2019) point out that infrastructure contributes to memorable and relaxing holidays for tourists who visit Thailand, especially for medical tourism, and if South Korea develops its infrastructure, it will become more competitive in Asia. In a similar tone, Kovačević et al. (2018) posit that tourism services such as food and accommodation services, and transport and taxi services are rated as top competitive determinants of South Banat district in Serbia.

In recent years, safety has been a major considering factor in tourist decision-making. The location and safety of a destination have become important indicators of competitiveness for emerging economies in the travel market (Kovačević et al., 2018). Empirical studies show that infrastructure is not a major concern for the developed economies, but a main driver of competitiveness in developing economies (Cvelbar et al., 2016). Without the infrastructure, tourist destinations in developing economies will not be attractive. However, most emerging destinations are confronted by inadequate general and tourism infrastructure. While a few emerging economies such as Singapore, UAE, Korea, Egypt, and Malaysia have leveraged their infrastructure development to achieve competitive advantage (Ganguli & Ebrahim, 2017; Khalifa, 2020; Michael et al., 2019; Nilashi et al., 2019), many emerging economies (including Serbia, Malawi, Hungary, Brazil, and Bulgaria) are battling with deficient accommodation capacities and structures, poor international connectedness, and poor general infrastructure among others (Kankhuni, 2020; Magombo et al., 2017; Pejanović et al., 2017), which decrease their competitive potential as emerging destinations.

Branding. Blain et al. (2005) posit that branding is important for creating memorable experiences. The competitiveness of emerging economies is driven by brands such as using slogans, and strategies aimed at attracting a specific market (Goffi et al., 2019; Kankhuni, 2020).

Successful emerging economies such as UAE, Korea, and Singapore have used strategic branding and marketing to gain a competitive advantage in the travel market (Ganguli & Ebrahim, 2017; Kim et al., 2019; Michael et al., 2019).

Different aspects of branding have been used to strengthen the unique image of emerging economies and to differentiate them from other destinations. Slogans have particularly been employed to differentiate destinations. Examples of these brand slogans include "The Warm Heart of Africa"-Malawi, "The Gateway to Africa"- Ghana, "Passion Made Possible" and "Singapore Medicine"- Singapore, "Definitely Dubai"-UAE, and "Asia's World City"-Hong Kong. Kankhuni (2020) argues that the use of a slogan is enough to set one destination apart from the other. However, empirical research has revealed that though communication of core brand names is good, symbols or logos also play a critical role in the competitiveness of emerging economies (Ganguli & Ebrahim, 2017). Logos also establish brands and enhance their competitiveness through differentiation. Emerging economies usually embark on ineffective marketing and branding campaigns. Poor marketing and branding in emerging economies are the results of a myriad of issues including, funding challenges, poor stakeholder collaborations (Kankhuni, 2020), low market orientation (Aqueveque & Bianchi, 2017), and poor business climate (Kovačević et al., 2018). There is, however, the recognition that if better branding and marketing strategies are established, emerging economies will improve their competitiveness in the travel market (Kankhuni, 2020).

Pricing. Price factors that include elements such as cost of transport services, tourist entertainment, accommodation, entry fees, visa fees, and food and beverage represent another important indicator of competitiveness, especially in emerging economies. Dwyer et al. (2000) assert that price differentials and the movement of exchange rate play a role in the attractiveness of a destination or otherwise. Evidence suggests that prices of tourism services and products form an essential component in the overall competitiveness of a destination because of the price sensitivity of tourism (Enderwick & Nagar, 2011) and may change over time (Plessis & Saayman, 2018). Xie and Tveteras (2020) are of the view that price competitiveness makes a destination more attractive. As emerging economies such as Malaysia, Singapore, and Thailand seldom emphasise price in their rapidly growing medical tourism, these markets basically compete on prices (Enderwick & Nagar, 2011). Emerging economies have the challenge of high prices of tourism and hospitality services, which are not competitive in comparison with developed destinations. Therefore, emerging economies, including Thailand and India, attempt to enhance their competitiveness using price competitive tourist packages (Enderwick & Nagar, 2011; Kim et al., 2019). In Africa, some economies pay attention to the price-demand dynamics to enhance their tourism competitive advantage. An example is South Africa, which capitalises on the law of demand such that it relies on the influence exerted on tourism products and package prices by the fluctuation in fuel and electricity prices. These prices improved in 2017. Subsequently, the increase in prices of tourism products and services was deemed acceptable by tourists (Plessis & Saayman, 2018). Other economies including Uganda, Ghana, and Nigeria can increase their attractiveness with price-competitive strategies (Ennin, 2017; Marcus et al., 2017), as they hold comparable attractive tourism resources, services, and infrastructure. However, most emerging economies are bedevilled by high cost of production which culminates into high prices, rendering such economies less price competitive in the ever-competitive tourism market.

Government Policy, Planning, and Development. Across emerging economies in Africa, Asia, America, and Oceania, the policies pursued by the majority of governments tend to influence the flow of tourists and the attractiveness of the destinations. Studies by Mishra et al. (2016) show that governance is a very critical determinant of the competitiveness of emerging economies. In line with this, Ganguli and Ebrahim (2017) and Nilashi et al. (2019) affirm that consistent and integrated policy structures that are executed among various stakeholders are critical to the competitiveness of emerging economies. Policies play an important role in attracting investment, improving tourism and general infrastructure, and facilitating travel to emerging destinations such as UAE, Singapore, and Malaysia. In Singapore, the government created the National Interagency made up of the Singapore Economic Development, International Enterprise Singapore, and the Singapore Tourism Board to encourage foreign direct investment, improve the capacity and people-oriented services to engage, and promote the local medical enterprises through international marketing and branding campaigns that meet the global medical tourism market. The government of Malaysia developed a policy geared towards the development of infrastructure including transportation, and hotels in readiness for tourism. In addition, the government of UAE established the Ajman Strategic Plan for Tourism 2015–2021 that ensures the provision of value-added services, enhance operational effectiveness, and sustainable growth of tourism. These policies positioned the destinations in a very competitive position over other destinations.

Policies that foster sustainable tourism development are important for the competitiveness of emerging tourist destinations because the sustainability of emerging destinations is an important element for their competitiveness (Goffi et al., 2019; Khalifa, 2020). Thus, strategic planning and practices guided by structured and focused policies improve the competitiveness of destinations in emerging economies (Goffi et al., 2019). It has been suggested by Lubbe et al. (2019) that having stricter policies that work improves competitiveness by reducing unsustainable activities that negatively affect the enjoyment, safety, and uncertainty in emerging economies. Similarly, Kankhuni (2020, p. 15) considers that to improve competitiveness, emerging economies must establish "a strong platform that encourages collaborations between government and the private sector and general citizenry". Policies can help to govern strategic leadership, planning, and practices to improve competitiveness (Khalifa, 2020). However, emerging economies are usually characterised by poor policy formation and implementation (Kankhuni, 2020). The political willpower to implement policies that have already been formulated is important to encourage adherence to the law and a single focus of all stakeholders in the given destination (Nilashi et al., 2019). An example is the outstanding competitive achievement attained by Singapore through international openness and economic liberalisation policies (Ganguli & Ebrahim, 2017). Such active demonstration of the willpower of governments encourages all stakeholders in a destination to have a unidirectional focus for the attainment of competitive advantage, yet emerging economies are known to have poor stakeholder collaboration at all levels of destination management.

Human Resource Development. The human factor of a destination is an important determinant in the competitiveness of emerging destinations. Nilashi et al. (2019) found that the human resources of emerging economies are among the critical elements in destination management for competitive readiness. The human factor of a destination, according to Nilashi et al. (2019), is essential for making visitors feel comfortable and enjoy their stay while operating and manipulating other technological and environmental factors to improve or maintain the destination's competitiveness. Ganguli and Ebrahim (2017) and Heath (2002) also affirm that having a large pool of skilled human resources with varied expertise stands an emerging destination a good chance to improve its competitiveness by delivering quality service. The human characteristics of a destination are linked to the image that the destination has. Therefore, human resource development is a crucial activity for every DMO.

Skilled employees are needed in every destination to offer the human touch to tourists' experience. The interactions and services rendered in tourism-related organisations are very important to the creation of memorable experiences among tourists such that the hiring of high-quality skilled service personnel is important to give a destination a competitive advantage over its competition. The UAE, for instance, has capitalised on the efficient labour market and highly trained and educated service personnel to gain a competitive edge in the tourism labour market (Michael et al., 2019). Unfortunately, most emerging economies lack proper skilled human resources.

Technology. Technology readiness is important for destinations in today's competitive world. Management of destinations through state of the art is very important to help develop innovative products that add uniqueness to the attraction and improve what the destination has to offer in terms of attractive resources. In addition, it facilitates stakeholder collaboration and helps to build a sustainable and competitive destination, which has become very important. Kankhuni (2020), Kovačević et al. (2018), Michael et al. (2019), and Nilashi et al. (2019) agree that emerging economies need innovative technologies with effective and efficient information systems to advance their competitiveness. These are more important regarding the fact that the technology readiness of a destination (Nilashi et al., 2019) will ensure proper responses to the demand condition, quality services, and memorable experiences (Aqueveque & Bianchi, 2017). Therefore, organisations in emerging economies are to continuously develop and/or implement innovative technologies that enhance information sharing, communication, productivity (Porter, 1990), and, in effect, the competitiveness of the destination. Currently, advancement in technology has made it possible for destinations to communicate and market their unique offerings and for potential tourists to plan, search for information, compare alternatives, buy travel packages, and experience travel and tourism by the touch of a button. Though internet penetration and access to ICT have increased, emerging destinations still have difficulties in capitalising on such opportunities.

Conclusion

For emerging destinations to remain truly competitive in the global tourism market, DMOs must create favourable images, develop their tourism resources, and ensure greater stakeholder collaboration. The fierce competition among tourist destinations around the world makes it necessary for destinations to adopt marketing activities that create favourable images, reduce perceived risks and costs, and create memorable travel experiences (Blain et al., 2005). This enhances the visitor-destination connection and the destination's image to differentiate it from others. Though emerging economies are endowed with natural and cultural resources, these alone do not determine the competitiveness of destinations (Ganguli & Ebrahim, 2017). The underdevelopment of natural and cultural resources, coupled with poor communication, and the lack of new product development and diversification in emerging economies as well as the high level of prices of tourism services and products render them less competitive (Ferreira & Perks, 2020; Kankhuni, 2020). Emerging economies must enhance their tourism resources with the requisite facilities, infrastructure, and superstructure. Moreover, collaboration among various stakeholders at the destination helps to determine a common goal, brand identity, and a more focused marketing agenda with a suitable market pricing strategy. Besides, the right regulatory policies should be formulated to guide the entire destination development and marketing process. More so, in this digital age, information search is an important element in tourist decision-making. Thus, for a destination to be competitive, it has to brand and market itself successfully, online and through other digital platforms. Finally, in the current global situation, travellers' safety (especially from COVID-19) is very important. Emerging destinations are to strive to be more competitive by reaching herd immunity through vaccination. With this, tourists will feel safe in emerging destinations and, hence, will consider such destinations for vacations and other tourism activities during and post COVID-19 pandemic recovery period.

References

Acemoglu, D., & Robinson, J. (2012). *Why nations fail: The origins of power, prosperity, and poverty*. Crown Business.

Alseiari, H. A. S. M., Khalifa, G. S. A., Al-Shibami, A. H., & Ghosh, A. (2019). Driving strategic leadership towards tourism sustainability in Abu Dhabi. *International Journal of Recent Technology and Engineering, 8*(4), 12137–12141.

Aqueveque, C., & Bianchi, C. (2017). Tourism destination competitiveness of Chile: A stakeholder perspective. *Tourism Planning & Development, 14*(4), 447–466.

Armenski, T., Gomezelj, O. D., Djurdjev, B., Đeri, L., & Dragin, A. (2011). Destination competitiveness: A challenging process for Serbia. *Human Geographies, 5*(*1*), 19–33.

Azarya, V. (2004). Globalization and international tourism in developing countries: Marginality as a commercial commodity. *Current Sociology, 52*(6), 949–967. https://doi.org/10.1177/0011392104046617

Azzopardi, E., & Nash, R. (2016). A framework for island destination competitiveness: Perspectives from the island of Malta. *Current Issues in Tourism, 19*(3), 253–281.

Bernini, C. (2009). Convention industry and destination clusters: Evidence from Italy. *Tourism Management, 30*(6), 878–889.

Blain, C., Levy, S. E., & Ritchie, J. R. B. (2005). Destination branding: Insights and practices from destination management organizations. *Journal of Travel Research, 43*(4), 328–338. https://doi.org/10.1177/0047287505274646

Buhalis, D. (2000). Marketing the competitive destination of the future. *Tourism Management, 21*, 97–116.

Candela, G., & Figini, P. (2012). *The economics of tourism destinations*. Springer.
Croes, R. (2010). Measuring and explaining competitiveness in the context of small island destinations. *Journal of Travel Research, 20*(10), 1–12.
Cronjé, D. F., & du Plessis, E. (2020). A review on tourism destination competitiveness. *Journal of Hospitality and Tourism Management, 45*, 256–265. https://doi.org/10.1016/j.jhtm.2020.06.012
Crouch, G. I. (2008). *Measuring tourism competitiveness: Research, theory and the WEF index*. Unpublished paper, La Trobe University.
Crouch, G. I. (2011). Destination competitiveness: An analysis of determinant attributes. *Journal of Travel Research, 50*(1), 27–45. https://doi.org/10.1177/0047287510362776
Crouch, G. I., & Ritchie, J. R. B. (1999). Tourism, competitiveness, and societal prosperity. *Journal of Business Research, 44*, 137–152.
Cvelbar, L. K., Dwyer, L., Koman, M., & Mihalič, T. (2016). Drivers of destination competitiveness in tourism: A global investigation. *Journal of Travel Research, 55*(8), 1041–1050.
d'Hauteserre, A. M. (2000). Lessons in managed destination competitiveness: The case of Foxwoods Casino Resort. *Tourism Management, 21*, 23–32.
Dwyer, L., & Forsyth, P. (2003). Inter-industry effects of tourism growth: Implications for destination managers. *Tourism Economics, 9*(2), 117–132.
Dwyer, L., Forsyth, P., & Rao, P. (2000). The price competitiveness of travel and tourism: A comparison of 19 destinations. *Tourism Management, 21*(1), 9–22.
Dwyer, L., & Kim, C. (2003). Destination competitiveness: Determinants and indicators. *Current Issues in Tourism, 6*(5), 369–414. https://doi.org/10.1080/13683500308667962
Echtner, C. M., & Ritchie, J. R. B. (2003). The meaning and measurement of destination image. *The Journal of Tourism Studies, 14*(1), 37–48.
Enderwick, P., & Nagar, S. (2011). The competitive challenge of emerging markets: The case of medical tourism. *International Journal of Emerging Markets, 6*(4), 329–350. https://doi.org/10.1108/17468801111170347
Ennin, I. B. (2017). Determinants and economic impact of international tourist arrivals in Ghana. Policy Brief. AERC CREA. Retrieved from: http://publication.aercafricalibrary.org/bitstream/handle/123456789/173/PB_29.pdf?sequence=1
Enright, M. J., & Newton, J. (2004). Tourism destination competitiveness: A quantitative approach. *Tourism Management, 25*(6), 777–788. https://doi.org/10.1016/j.tourman.2004.06.008

Ferreira, D., & Perks, S. (2020). A dimensional framework of tourism indicators influencing destination competitiveness. *African Journal of Hospitality, Tourism and Leisure, 9*(3), 1–21. https://doi.org/10.46222/ajhtl.19770720-1

Gallego, I., & Font, X. (2020). Changes in air passenger demand as a result of the COVID-19 crisis: Using big data to inform tourism policy. *Journal of Sustainable Tourism, 29*, 1–20.

Ganguli, S., & Ebrahim, A. H. (2017). A qualitative analysis of Singapore's medical tourism competitiveness. *Tourism Management Perspectives, 21*, 74–84. https://doi.org/10.1016/j.tmp.2016.12.002

Goffi, G., Cucculelli, M., & Masiero, L. (2019). Fostering tourism destination competitiveness in developing countries: The role of sustainability. *Journal of Cleaner Production, 209*, 101–115. https://doi.org/10.1016/j.jclepro.2018.10.208

Hassan, S. S. (2000). Determinants of market competitiveness in an environmentally sustainable tourism industry. *Journal of Travel Research, 38*(3), 239–245. https://doi.org/10.1177/004728750003800305

Heath, E. (2002). Towards a model to enhance Africa's sustainable tourism competitiveness. *Journal of Public Administration, 37*(Special issue 1), 327–353.

Ibrahim, Z., Zahari, M. S., Sulaiman, M., Othman, Z., & Jusoff, K. (2009). *Travelling pattern and preferences of the Arab tourists in Malaysian hotels*. Universiti Teknologi MARA.

Kankhuni, Z. (2020). Tourism destination branding in Malawi: A supply-side perspective. *African Journal of Hospitality, Tourism and Leisure, 9*(1), 1–21. http://www.ajhtl.com

Khalifa, G. S. A. (2020). Factors affecting tourism organization competitiveness: Implications for the Egyptian tourism industry. *African Journal of Hospitality, Tourism and Leisure, 9*(3), 116–130. https://doi.org/10.46222/ajhtl.19770720-8

Kim, S., Arcodia, C., & Kim, I. (2019). Critical success factors of medical tourism: The case of South Korea. *International Journal of Environmental Research and Public Health, 16*(24) https://doi.org/10.3390/ijerph16244964

Kovačević, N. D., Kovačević, L., Stankov, U., Dragićević, V., & Miletić, A. (2018). Applying destination competitiveness model to strategic tourism development of small destinations: The case of South Banat district. *Journal of Destination Marketing & Management, 8*, 114–124.

Kozak, M., Baloğlu, S., & Bahar, O. (2009). Measuring destination competitiveness: Multiple destinations versus multiple nationalities. *Journal of Hospitality Marketing & Management, 19*(1), 56–71.

Kozak, M., & Rimmington, M. (1999). Measuring tourist destination competitiveness: Conceptual considerations and empirical findings. *International Journal of Hospitality Management, 18*(3), 273–283. https://doi.org/10.1016/S0278-4319(99)00034-1

Krešić, D., & Prebežac, D. (2011). Index of destination attractiveness as a tool for destination attractiveness assessment. *Tourism - An International Interdisciplinary Journal, 59*(4), 497–517.

Krugman, P. (1994). Competitiveness: A dangerous obsession. *Foreign Affairs, 73*(2), 28–44.

Laws, E., Richins, H., Agrusa, J., & Scott, N. (2011). *Tourist destination governance: Practice, theory and issues*. CAB International.

Lubbe, B. A., du Preez, E. A., Douglas, A., & Fairer-Wessels, F. (2019). The impact of rhino poaching on tourist experiences and future visitation to national parks in South Africa. *Current Issues in Tourism, 22*, 8–15. https://doi.org/10.1080/13683500.2017.1343807

Magombo, A., Rogerson, C. M., & Rogerson, J. M. (2017). Accommodation services for competitive tourism in sub-Saharan Africa: Historical evidence from Malawi. *Bulletin of Geography. Socio-Economic Series, 38*(38), 73–92. https://doi.org/10.1515/bog-2017-0035

Marcus, A. O., Rejoice, O. E., Kenechi, A. E., & Luke, N. N. (2017). A study of relationship between tourism price competitiveness and international tourist arrivals: A comparative study of Nigeria and selected African countries. Development (UNCTAD), 1.

Michael, N., Reisinger, Y., & Hayes, J. P. (2019). The UAE's tourism competitiveness: A business perspective. *Tourism Management Perspectives, 30*, 53–64. https://doi.org/10.1016/j.tmp.2019.02.002

Mishra, A. K., Rao, G., Monga, A., & Vishwanath, B. (2016). Assessing competitiveness in emerging Asian economies: Role of governance and infrastructure and lessons for India. *Emerging Economy Studies, 2*(1), 72–90. https://doi.org/10.1177/2394901515627732

Mitchell, J., & Faal, J. (2007). Holiday package tourism and the poor in the Gambia. *Development Southern Africa, 24*(3), 445–464.

Mulec, I., & Wise, N. (2013). Indicating the competitiveness of Serbia's Vojvodina Region as an emerging tourism destination. *Tourism Management Perspectives, 8*, 68–79.

Nilashi, M., Samad, S., Manaf, A. A., Ahmadi, H., Rashid, T. A., Munshi, A., Almukadi, W., Ibrahim, O., & Ahmed, O. H. (2019). Factors influencing medical tourism adoption in Malaysia: A DEMATEL-Fuzzy TOPSIS approach. *Computers and Industrial Engineering, 137*. https://doi.org/10.1016/j.cie.2019.106005

Omerzel, D. G. (2006). Competitiveness of Slovenia as a tourist destination. *Managing Global Transitions, 4*(2), 167–189.

Papp, Z., & Raffay, Á. (2011). Factors influencing the tourism competitiveness of former socialist countries. *Human Geographies-Journal of Studies and Research in Human Geography, 2*, 21–30. https://doi.org/10.5719/hgeo.2011.52.21

Pejanović, R., Demirović, D., Glavaš-Trbić, D., Maksimović, G., & Tomaš-Simin, M. (2017). Clusters as a factor of competitiveness of rural tourism destinations in the Danube region of the Republic of Serbia. *Tourism Economics, 23*(2), 475–482. https://doi.org/10.5367/te.2015.0509

Plessis, E. D., & Saayman, M. (2018). Aspects contributing to tourism price competitiveness of South Africa. *Tourism Economics, 24*(2), 146–156.

Porter, M., Sachs, J., & McArthur, J. (2001). *Executive summary: Competitiveness and stages of economic development.* In World Economic Forum The Global Competitiveness Report 2001–2002.

Porter, M. E. (1990). The competitive advantage of nations. *Harvard Business Review, 68*(2), 73–93.

Qiu, R. T., Wu, D. C., Dropsy, V., Petit, S., Pratt, S., & Ohe, Y. (2021). Visitor arrivals forecasts amid COVID-19: A perspective from the Asia and Pacific team. *Annals of Tourism Research, 88*, 103155.

Rey-Maquieira, J., & Ramos, V. (2006). Deatination competitiveness. In J. Jafari & H. Xiao (Eds.), *Encyclopedia of tourism* (pp. 244–245). Springer. https://doi.org/10.1007/978-3-319-01384-8_54

Ritchie, J. R. B., & Crouch, G. I. (2003). *The competitive destination: A sustainable tourism perspective*. CABI.

Salinas Fernández, J. A., Serdeira Azevedo, P., Martín Martín, J. M., & Rodríguez Martín, J. A. (2020). Determinants of tourism destination competitiveness in the countries most visited by international tourists: Proposal of a synthetic index. *Tourism Management Perspectives, 33*. https://doi.org/10.1016/j.tmp.2019.100582

Sanches-Pereira, A., Onguglo, B., Pacini, H., Gómez, M. F., Teixeira Coelho, S., & Muwanga, M. K. (2017). Fostering local sustainable development in Tanzania by enhancing linkages between tourism and small-scale agriculture. *Journal of Cleaner Production, 162*, 1567–1581.

Swart, K., & Maralack, D. (2020). COVID-19 and the cancellation of the 2020 two oceans marathon, Cape Town, South Africa. *Sport in Society, 23*(11), 1736–1752.

Teixeira, S. J., & Ferreira, J. J. M. (2019). Entrepreneurial artisan products as regional tourism competitiveness. *International Journal of Entrepreneurial Behavior & Research, 25*(4), 652–673. https://doi.org/10.1108/IJEBR-01-2018-0023

U.S. Department of Health & Human Services. (2021, May 13). *COVID data traker*. https://covid.cdc.gov/covid-data-tracker/#vaccinations

UNWTO. (2011). *Tourism towards 2030: Global overview*.

Waheeduzzaman, A. N. M. (2011). Competitiveness and convergence in G7 and emerging markets. *Competitiveness Review, 21*(2), 110–128.

WEF. (2015a). *The travel & tourism competitiveness report 2017*. WEF.

WEF. (2015b). The *travel & tourism competitiveness index 2015*: T&T as a resilient contribution to national development. http://www3.weforum.org/-docs/TT15/Zeland. http://www.fama2.us.es:8080/turismo/turismonet1/economia%20del%20turismo/demanda%20turistica/DESTINATION%20COMPETITIVENESS%20A%MODEL%AND%20DETERMINANTS

WEF. (2017). *The travel & tourism competitiveness report 2017*. WEF.

WEF. (2019a). *The travel & tourism competitiveness report 2017*. WEF.

World Economic Forum (WEF). (2012). *The ASEAN travel & tourism competitiveness report 2012*. WEF.

Xie, J., & Tveterås, S. (2020). Economic decline and the birth of a tourist nation. *Scandinavian Journal of Hospitality and Tourism, 20*(1), 49–67. https://doi.org/10.1080/15022250.2020.1719882

Part II

Travel Behaviour and ICT Applications in Destination Marketing

5

Travel to Emerging Tourist Destinations: Motivations and Decision-Making Processes

Benjamin Appiah Osei

Introduction

The most basic decisions made by tourists include whether to travel, where to travel, when to travel, how long to stay, what to do, and how much to spend. Although tourism stakeholders and destination marketers have an interest in how tourists come out with all these decisions, the most significant of these issues relates to where to travel, which is the choice of destination. Similar to tourist destinations in advanced economies, emerging tourist destinations also encompass places where tourists travel to and engage in activities that are dependent on nature, culture, and other resources during the period of stay at the destination.

Aside from being tourism products that offer an integrated experience to tourists, destinations also have images that are mainly interpreted by their consumers, based on a broad range of determinants. Hence, what

B. A. Osei (✉)
School of Hospitality, Tourism and Events, Taylor's University, Selangor, Malaysia

I. Mensah et al. (eds.), *Marketing Tourist Destinations in Emerging Economies*, Palgrave Studies of Marketing in Emerging Economies,
https://doi.org/10.1007/978-3-030-83711-2_5

motivates people to embark on trips and the decision to undertake such trips should be issues of concern for destination marketers and stakeholders. In this regard, this chapter begins with an explanation of the factors that influence travel to emerging destinations. This is followed by a detailed theoretical and empirical review of the motivations for travel to emerging destinations. Again, the concepts of travel decision-making process and how they could be applied to emerging destinations are explained. Also, the relationship between travel motivations and decision-making is examined. Finally, the chapter concludes by exploring the implications of travel motivations and decision-making for the marketing of emerging destinations.

Factors that Influence Travel to Emerging Destinations

The choice of destination has always been a central issue in the travel behaviour literature (Dellaert et al., 1998; Papatheodorou, 2006; Van Vuuren & Slabbert, 2011), with diverse factors influencing tourists' decision-making or purchase behaviour. It is noteworthy that tourists' travel decisions consist of complex multifaceted and interrelated decisions (Nejati & Mohamed, 2014). Travel choices and preferences are dependent on subjective perceptions of the individual and vary from one person to another. Over the years, scholars and researchers have sought to identify the factors that influence travel to a particular destination.

These factors include travel motivations, tourists' attitudes and intentions, environmental factors, safety and security, and value for money (Nejati & Mohamed, 2014; Seyidov & Adomaitienė, 2017; Van Vuuren & Slabbert, 2011). The tourism literature on the factors influencing travel, to especially emerging destinations, is often grouped under two major caveats. The first one involves factors that are grouped under internal and external factors (Hill, 2000; Horner & Swarbrooke, 1996; Hsu et al., 2009; Schmoll, 1977; Um & Crompton, 1990). The second group consists of social, cultural, personal, psychological, and environmental factors (Kotler et al., 2006; Seyidov & Adomaitienė, 2017; Yoo et al., 2018).

For the first group of factors that influence travel, various scholars have highlighted different internal and external factors. For instance, Schmoll (1977), in his theory of travel decision, explained that tourists' purchase intentions are as a result of the interaction of four fields of influence that are both internal and external to the tourist. These fields include characteristic features of the service destination (e.g., range of attractions and value), travel stimuli (e.g., advertising and travel literature), external variable (e.g., confidence in travel trade intermediaries), and personal determinants of travel behaviour (e.g., personality features, attitudes, and values).

Um and Crompton (1990) explained the internal and external factors that influence travel in their framework of travel destination choice process. The authors highlighted that the internal factors encompass the socio-psychological make-up of a potential tourist. These include personal characteristics (e.g., socio-demographics, lifestyle, personality, and situational factors), motives, values, and attitudes. On the other hand, the external factors consist of the sum of social interactions and marketing communications to which a potential pleasure traveller is exposed. In addition to these two factors, Um and Crompton also added another set of factors that influence travel, which are cognitive constructs. They explained cognitive constructs as an integration of the internal and external inputs into the awareness set and the evoked set of the destinations.

Hill (2000) also explained that travellers' decisions are influenced by personal or internal characteristics (e.g., motivation, perceptual encoding, use of memory, and decision rules), as well as external forces (e.g., culture, family, lifestyle, and situational variables). Horner and Swarbrooke (1996) gave examples of the internal factors influencing holiday decision as personal motives, personality, disposable income, health, family commitments, work commitments, past experience, hobbies and interests, existing knowledge of potential holiday destination, lifestyles, attitudes, opinions, and perceptions. For external factors influencing travel decision, Horner and Swarbrooke gave examples as availability of suitable products, advice of travel agents, political restrictions on travel, visa and health regulations, information obtained from destination marketers, word-of-mouth recommendation of friends and family members, special promotions and offers from tourism organisations, and climate of the destination regions. Hsu et al. (2009) also presented examples of internal factors that influence travel decision as desire for escape, rest, relaxation, prestige, health fitness, adventure, and social interaction, whereas external factors are based on the attractiveness of the destination, which include tangible resources, and tourists' perceptions and expectations.

For the second group, the travel decision or behaviour of tourists is mainly influenced by social, cultural, personal, psychological, and environmental factors. These factors contribute to the choice of destination, coupled with the services and products that will be purchased by the tourists (Kotler et al., 2006; Seyidov & Adomaitienė, 2017). The social factors simply refer to the influential nature of groups, families, individual roles, and status on travel to a particular destination. The cultural factors, on the other hand, encompass all elements of culture, sub-culture, and social class which influence tourists' choice of a destination. Personal factors that influence travel decisions include age, life cycle, occupation, economic situation, personality, and lifestyle. Psychological factors could be subdivided into other factors such as motivation, perception, learning, beliefs, and attitudes, which influence tourists' choice of destination.

Lastly, environmental features include destination amenities, image, climate, cleanliness, and safety that influence travel behaviour. Other factors influencing travel to an emerging economy like Malaysia are the allocation of money and cost, safety, family togetherness, access to the destination, and new environment (Segumpan et al., 2010). Also, Nejati and Mohamed (2014), in their study, identified three major destination factors that influence travel decisions of tourists to Penang Island: (a) quality, (b) distance and value, and (c) uniqueness. In view of this, it is worth noting that the literature on the factors influencing travel is often discussed in line with motivations that affect tourists' needs and behaviour. These travel motivations are further discussed in the next sub-topic.

Travel Motivation

Motivation initiates when a person is on a quest to satisfy a need and is driven towards certain activities that the person must execute to derive satisfaction. Van Vuuren and Slabbert (2011, p. 296) defined motivation as "the inner state of a person, or certain needs and wants of a person, which forces them to act or behave in a specific way and thus sustaining human behaviour and energy levels of the human body". In tourism and hospitality, motivation is considered a psychological or physical need that awakens, synchronises, and directs an individual's behaviour and activity (Khan et al., 2017). In the tourism literature, travel motivation is regarded as one of the most salient psychological factors influencing tourist decision-making and behaviour (Seyidov & Adomaitienė, 2017). In line with this, Mutanga et al. (2017, p. 2) defined tourism motivation as "the set of needs which influence a person to partake in a tourism activity".

These gravitational actors and destination sources act as factors that influence tourists' decision to choose a particular destination. Nonetheless, the challenge with changing needs of tourists and their complex behaviour needs to be highlighted when addressing travel motivations. Practically, due to the continuous adaptations of human needs and complex behavioural patterns, it still remains difficult to fully comprehend the motivations behind people's travel preferences and destination choices (Khan et al., 2017). Again, it is worth noting that tourists' motivations are subjective, in that what motivates one tourist to travel is likely to be different for another tourist. Hence, understanding motivations from tourists' perspective is very key. Following this, theories underpinning travel motivation as well as factors that motivate tourists to travel to emerging destinations are further discussed.

Theories of Travel Motivation

Traditional motivation theories in social psychology and sociology have contributed to answering the basic question, why people travel. Scholars have highlighted the importance of such theories in the tourism literature (Jang & Cai, 2002; Jang et al., 2009; Van Vuuren & Slabbert, 2011). On the other hand, others have also addressed this phenomenon by focusing on the psychological aspects of travel and tourism. With this, scholars argue that touristic activities are in nature embedded in psychological traits; hence, tourists tend to seek these psychological rewards associated with travel (Park et al., 2019). For the social psychology theories of motivation, one cannot talk about travel motivations and neglect the famous

Maslow's hierarchy of needs theory. The theory of needs by Maslow (1954) is one of the most frequently used frameworks in explaining the phenomenon of motivation across various academic disciplines, including tourism. To date, many tourism studies on travel motivation base their theoretical background on Maslow's five-stage hierarchy of needs theory, which categorises human needs into a set hierarchy. Maslow (1954) highlighted five sets of hierarchical order goals of increasing motivational relevance. These include physiological, safety, social, self-esteem, and self-actualisation needs.

Subsequently, in 1977, Dann (1977) proposed a theory in the tourism motivation literature. This theory is popularly referred to as the "push" and "pull" factors that motivate people to travel. Dann's theory is considered as one of the most widely accepted theories of travel motivation in the tourism literature (Goossens, 2000; Jang & Cai, 2002; Jang et al., 2009). The push factors refer to the socio-psychological needs which are internal to the individual and inspire a desire for the individual to travel (Khan et al., 2017). On the other hand, the pull factors refer to the specific destination uniqueness and attributes that are external to the individual and affect where, how, and when the individual decides to travel (Seyidov & Adomaitienė, 2017).

Following Dann's (1977) push and pull theory, Crompton (1979) also theorised motivations for embarking on pleasure vacation. Crompton (1979) identified nine motives of tourists who travel for pleasure. Specifically, the scholar argued that tourists' motivation encapsulates various components which include social, psychological, cultural, economic, environmental, and/or situational determinants, all of which influence the decision to travel (Crompton, 1979; Park et al., 2019). Iso-Ahola, in 1982, proposed a theoretical framework of forces of travel motivation. These forces involved two major components of travel or leisure motives. These dimensions of travel forces include seeking/approach/reward and escapism/avoidance (Iso-Ahola, 1982; Park et al., 2019). Iso-Ahola (1982) argued that tourists' motivation and behaviour to travel or undertake a leisure activity result from these two psychological dimensions (i.e., seeking and escaping).

Pearce (1988) developed a career ladder model that sought to explain travel motivation, grounded in Maslow's hierarchy of needs. Specifically, the travel career ladder (TCL) constituted five steps that motivate tourists to travel. According to the TCL, Pearce explained that tourists' travel motivation aggregates and/or amends throughout their lives, vis-à-vis their travel engagements. In other words, tourists differ with regard to their travel career level, due to their motives for travel (Jang & Cai, 2002).

Although the theories and models are all explaining tourism motivation concepts and rooted in changes in tourists' behaviour regarding travel over time, these theories are often applied independently. Additionally, given the complexities with today's tourists and tourism environment, an investigation solely focusing on only the psychological aspects with travel motivation and decision-making will not entirely address the question, why tourists travel or make their destination choices. Practically, in these times of COVID-19 pandemic, today's tourists are faced with more sophisticated and complex factors affecting their engagement with tourism. For instance, it has been a long-existing knowledge in the tourism literature that since the 1990s, modern tourists are involved with varying needs, wants, expectations, and/or roles regarding their behaviours (Cohen, 1972).

Motives for Travel to Emerging Destinations

Previous studies have applied the push and pull theory and identified some factors that explain tourists' motives for travelling to emerging destinations. Nassar et al. (2015) identified push factors by Kuwaiti nationals as spending time with family and loved ones, as well as travelling to relax with nature. On the other hand, other pull factors were low transportation cost, price, travel distance, tourist attractions, weather conditions, product and service quality, shopping opportunities, destination image, security, safety, and accessibility.

Mutanga et al. (2017) also identified internal motives (push factors) of tourists to national parks in Zimbabwe as "need to escape, self-discovery, relaxation, prestige, challenge, income and adventure", whereas external motives (pull factors) included "scenic beauty, climate, history, culture, and sports" (p. 2). Also, Park et al. (2019, p. 65) highlighted push factors that motivate tourists to travel to Langkawi, Malaysia as "taste, fatigue, relaxation, social belongingness, exploration and evaluation of self, prestige, enrichment of relationships, and enabling of social interaction". For the pull factors, the researchers identified "cost of travel, income, expenditure, inflation, employability, destination distance, tourist location, seasons, and flora and fauna" as motives for travel (Park et al., 2019, p. 65).

Other studies also identified other factors that influence travel to emerging destinations. For instance, Sellick (2004) identified four travel motives that influence Australian senior tourists to embark on travel. These motives include discovery and self-enhancement, reluctant travellers, enthusiastic connectors, and nostalgic travellers. McNamara and Prideaux (2010) investigated the factors that influence solo young women (with the majority aged under 20) to travel to new destinations. The study revealed that the natural environment, climate, rest, meeting new people, relaxation, wildlife, and visiting friends and relatives were the motives for their travel. Again, a study by Chen et al. (2014) categorised Chinese tourists based on their travel motives. These included social seekers, self-actualisers, and destination experiencers. Similarly, a study by Van Vuuren and Slabbert (2011) explored the travel motivations of tourists who visited resorts in South Africa. They found that the motivations associated with travel to the resort were "recreational activities" (e.g., entertainment, fun, family time, and to do something out of the ordinary), "resting and relaxation" (e.g., breaking away from routine and pressure in the everyday environment, to escape from their surrounding environment), "personal values" (e.g., spending time with family, children and friends, travel due to the level of security it offers, and travelling with regard to their status), "social experiences" (e.g., making new friends, meeting new people, to gain more knowledge about the country, and travelling for social interaction), as well as "enriching and learning experiences" (i.e. natural surroundings of the resort, discovering new places, cultural activities, improvement of quality of life and excitement) (p. 298).

Travel Decision-Making

Tourists' behaviour refers to their attitude, decisions, activities, experiences, and ideas used to purchase, search, and evaluate tourism products and services that satisfy their needs (Choi et al., 2011; Seyidov & Adomaitienė, 2017; Van Vuuren & Slabbert, 2011). In other words, it refers to how tourists form their decision to spend their discretionary time and money on tourism products and services. Usually, to understand tourist behaviour and decision-making, the "choice set" concept has often been applied to explain the process. According to this classic approach, during an individual's daily decision-making, information is searched and analysed with the aim of selecting the optimal solution from a different range of alternatives (Edwards, 1954; Sirakaya & Woodside, 2005; Von Neumann & Morgenstern, 1944).

Choice is the main stage that is directly linked to an action, and it deals with moving from motivation to the buying process (Moutinho, 1987). This concept is well known in the tourism destination choice literature, where it is purported that tourists seek to make optimal selection among a range of alternatives regarding their travel decision-making (Jafari, 2013; Nejati & Mohamed, 2014). However, other scholars have questioned this approach, owing to the challenges caused by time constraints. According to these researchers, a more realistic concept to explain decision-making would be "bounded rationality" (March & Simon, 1958; Simon, 1955). The concept of bounded rationality indicates that tourists render decisions that they see to be good and offer them satisfaction and not just making optimal choices (Sirakaya & Woodside, 2005). This approach to decision-making is "less cognitively bound and more context-dependent; and considers a socially constructed approach towards tourist decision-making" (Nejati & Mohamed, 2014, p. 109).

Travel decision occurs when a solution to a need/problem is selected for satisfaction. Travel decision-making process is considered a very complex, dynamic, multifaceted course, and information-intensive processing that involves different interrelated stages or sub-decisions and concepts (Osei & Abenyin, 2016; Osei et al., 2018; Seyidov & Adomaitienė, 2017). Practically, the travel decision-making process involves a series of information searches as a result of the composite and experiential nature of the tourism service and products (Osei & Abenyin, 2016; Rathonyi, 2013). Also, this process is often broken into smaller and familiar elements or factors by tourists, such as accommodation, transportation, and feeding (Nejati & Mohamed, 2014). Stages of the travel decision-making process have been proposed by scholars in the tourism field. These are further highlighted in the next sub-topic.

Models of Travel Decision-Making

Generally, much of the tourist decision-making literature has been based on the classical buyer behaviour theory, especially the decision-making model by Engel-Kollat-Blackwell. In this particular model, decision-making is thought to evolve in five sequential stages or steps (Engel et al., 1978). These include problem recognition, information search, evaluation of alternatives and decision, purchase, and post-purchase evaluation (Engel et al., 1978; Osei & Abenyin, 2016). This can be seen in Fig. 5.1:

Drawing from this model, a number of travel decision-making models have been proposed to explain the travel decision-making processes of tourists or to unravel the general complexities with the internal and external factors influencing travel decision-making. These include well-known decision-making models in tourism by Van Raaij and Francken (1984), Moutinho (1987), Woodside and Lysonski (1989), and Um and Crompton (1990).

These traditional travel decision-making models have been commonly applied in many tourism studies. As a matter of fact, they support the notion

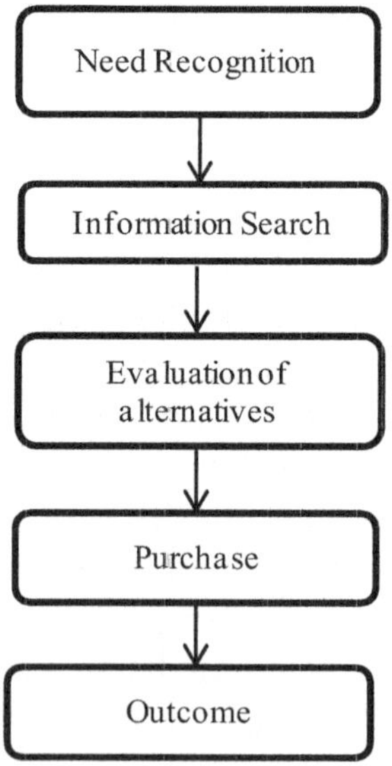

Fig. 5.1 Model of consumer decision-making. (Source: Engel et al., 1978)

that travel decision-making is a contingent and sequential process, and tourists follow a hierarchical approach of narrowing down choices amidst alternatives (Jeng & Fesenmaier, 2002; Sirakaya & Woodside, 2005). However, some researchers have critiqued the monolithic and deterministic view of these models of travel decision-making (Decrop, 1999; Hyde, 2004; Sirakaya & Woodside, 2005). These sceptics have argued that these models portray tourists as functional or rational decision-makers who make optimal choices out of alternatives (Hyde, 2004). Also, some concerns have been raised about the focus of these models on only destination choice, neglecting other multifaceted decisions during the entire travel planning (Fesenmaier & Jeng, 2000; Sirakaya & Woodside, 2005).

Travel Motivation and Decision-Making Nexus

It is an established fact that motivation initiates the entire travel decision-making process of tourists (Van Vuuren & Slabbert, 2011). Motivation occurs when a need ought to be satisfied. The outcome of the need is the direct representation of one's behaviour. Hence, the end results of this motivation (i.e., satisfaction of this need) will lead to an action or decision. Motivation is an important psychological variable in relation to tourists' decision-making and their travel satisfaction (Nejati & Mohamed, 2014; Yoo et al., 2018). This shows that there is a direct association between travel motivation and travel decision-making. This relates to where to visit and activities to undertake at the particular destination.

Furthermore, there have been studies to examine the influence of travel motivation on the decision-making process, market segmentation, and destination choice to emerging destinations. These studies have shown that once the motivation to travel is initiated, tourists begin to search and gather information in relation to their trips to emerging destinations, such as Malaysia (Battour et al., 2014) and Ghana (Osei & Abenyin, 2016). Other scholars have also argued that motivation is the key variable that stimulates the decision process before the actual travel, which is synonymous with what others have termed as travel intention (Jang et al., 2009; Mansfeld, 1992).

These aforementioned studies also proved that travel motivation also influences destination choice, in the sense that pull factors are triggered when tourists think about certain destinations and the range of activities they offer. Moreover, Jang and Cai (2002) identified a strong causal relationship between motivation and destination choice in Asia, indicating that both push and pull factors influence tourists' decision-making regarding their choice of destination. Jang et al. (2009), using Taiwanese seniors, also elucidated that focusing on the strengths in tourists' push and pull motivation factors provides an added benefit of understanding their travel behaviour, as well as providing a competitive advantage for a destination.

In essence, travel motivation forms the core of travel decision-making and behaviour in the tourism literature (Park et al., 2019). This is because in order to understand, explain, or theorise travel decision-making, one has to essentially explore travel motivation. Hence, from such a premise, it can be conclusively established that tourists' motivations have a strong link with their decision-making. Hence, travel motivation significantly influences tourists' perceptions, involvement, attitudes, and decision-making to emerging destinations (Jang & Cai, 2002; Khan et al., 2017; Park et al., 2019). Knowledge of this will help emerging destination marketers to enhance their product offerings.

Implications for the Marketing of Emerging Destinations

Destination marketing is about how a destination can sell its product offerings locally and internationally (Nejati & Mohamed, 2014). As already noted, it is of prime importance for destination marketers and industry service providers to understand the key factors that motivate tourists and influence their decision to travel to a particular destination. Knowledge of these factors provides key information for overall destination management and marketing. In the marketing literature, it is often claimed that the starting point for all market planning is the ability to understand customers' needs, expectations, and behaviour (Nassar et al.,

2015). Specifically, understanding tourists' motivations and behaviour goes a long way to assist destination marketers to adopt and develop the right strategies, effective promotional programmes, enhanced destination image, customised service delivery, and competitive advantage (Martin & Woodside, 2012; Nassar et al., 2015).

Market Planning

Considering the competitive tourism environment today, it is necessary for destination marketers to position their destinations on the minds of tourists. This should be in relation to travel motivations and decision-making. Hence, destination marketing, development, and management activities should be strategically planned by taking into consideration these factors, since they uniquely motivate travellers to decide on a destination. Furthermore, it is important for destination marketers to ascertain the kinds of information that are presented regarding tourists' decision-making in the multistage sequence of the travel-planning process (Choi et al., 2011).

For instance, Dellaert et al. (2014) illustrated that when tourists are faced with a decision challenge, they automatically create a mental replicate of the alternatives and the context that warrants the decision. Thus, knowledge of these key determinants of tourists' decision-making will offer destination marketers information on choice modelling, as well as factors that influence tourists' cognitive and psychological representations with decision-making. Therefore, identifying these key factors will go a long way to help in strategic planning and implementation of more effective destination strategies (Nejati & Mohamed, 2014). For example, a study by Nejati and Mohamed (2014) identified "quality", "uniqueness", and "distance and value" as key factors of travel decisions to Penang Island in Malaysia. This provided adequate knowledge and understanding to destination marketers about tourists' behaviour, further leading to the development and implementation of key policies that increased tourism demand in the area.

Promotion and Advertising

Jang and Cai (2002) explained that it is important for destination marketers to establish a strong fit between tourists' motivations and their destination attributes through effective promotional and advertising programmes. A study by Nassar et al. (2015) found that tourists were likely to be more responsive to promotional and advertising messages that focused on their reasons for travel, such as religion, visiting friends and relatives, shopping, recreation, and wellness. For instance, they identified that some tourists are likely to be drawn to destinations through cognitive images that conform to their expectations, hospitable destinations, as well as well-known destinations that have been accorded with a good international image. Hence, destination marketers should advertise through mass media, the popular destination images that appeal to tourists' cognitive and rational reasoning.

On the other hand, Nassar et al. (2015) also identified tourists who were motivated by affective images, especially images that make them proud of their identity, fulfil their needs, enhance their self-awareness, and grant them a sense of belongingness at the destination. With these kinds of tourists, destination marketers and promoters should apply "more sensory-stimulating media and develop more emotional messages to create an awareness of the traveller who makes destination choices using emotion rather than knowledge and reason" (Nassar et al., 2015, p. 45). The engagement of their emotions and stimulation of their senses via these advertising messages will help to better meet the internal desires of the tourists (e.g., for enhanced self-esteem). Likewise, it will develop positive impressions and propelling memories of the destination, thereby creating a strong edge to visit or revisit. Additionally, by concentrating on tourists' feelings, destination marketers will be able to connect effectively with tourists.

Enhanced Destination Image

Understanding tourists' motivations and decision-making allows emerging destination marketers to have a general idea and analysis of the tourism demand at the destination (Battour et al., 2014; Nejati & Mohamed,

2014). Aside from advertisement, meeting tourists' needs also leads to enhanced destination image. For instance, knowledge of tourism demand enables destination marketers to know their potential visitors' demand of them at the destination, hence, the need to keep their destination in shape. Thus, enhancing the quality of the destination also leads to continuous improvement in tourism activities at the destination. For instance, in modelling Hawaiian tourists' behaviour, Martin and Woodside (2012) highlighted that meeting tourists' demands and satisfying their needs were very crucial in influencing Hawaii's destination image. Furthermore, this went a long way to spread positive word of mouth about Hawaii, thereby enhancing the image of the destination in the minds of future tourists.

Customised Service Delivery and Competitive Advantage

For effective destination marketing and service delivery, destination marketers must understand what motivates individuals to travel and what destination attributes are essential for tourists when choosing a destination (Khan et al., 2017). This will go a long way to help destination marketers to meet the increasing demand and offer tailor-made services to tourists. Specifically, when the factors influencing the choice of a destination's tourism products and services by tourists are clear to destination marketers, they will be able to provide customised services to suit tourists' needs via appropriate marketing strategies (Battour et al., 2014). For instance, a study by Mutanga et al. (2017) on travel motivation to national parks in Zimbabwe found that motives differ from tourist to tourist, and a tourist could be motivated by more than one factor at a particular time. They further explained that understanding these different factors assisted in tourists' segmentation into different markets, as well as providing distinct products for tourists based on their needs. In short, knowledge about travel behaviour helps in destination marketing and product development, which in turn offers a competitive advantage to that particular destination.

Conclusion

Tourism continually serves as a major foreign exchange earner and a contributor to a destination's gross domestic product. Hence, it is very important for stakeholders to understand why tourists travel and their choice of destination. Also, today's tourism has become a sophisticated and specialised business commodity, with well-informed and demanding tourists (Yoo et al., 2018). Emerging destination marketers should put in much effort at identifying and responding to this diversified commodity and the tourists' changing demands. One of such means, as highlighted earlier, is to understand the motivations behind tourists' decision to travel, in order to meet their needs. Again, understanding tourists' travel decision-making process to emerging destinations is the key towards tourists' segmentation and customised service delivery. Efforts by destination marketers to understand these issues will assist them implement strategies that will enhance the image of the destination.

References

Battour, M., Ismail, M. N., Battor, M., & Awais, M. (2014). Islamic tourism: An empirical examination of travel motivation and satisfaction in Malaysia. *Current Issues in Tourism, 20*(1), 50–67.

Chen, G., Bao, J., & Huang, S. S. (2014). Segmenting Chinese backpackers by travel motivations. *International Journal of Tourism Research, 16*(4), 355–367.

Choi, S., Lehto, X. Y., Morrison, A. M., & Jang, S. (Shawn). (2011). Structure of travel planning processes and information use patterns. *Journal of Travel Research, 51*(1), 26–40.

Cohen, E. (1972). Toward a sociology of international tourism. *Social Research, 39*(1), 164–182.

Crompton, J. L. (1979). Motivations for pleasure vacation. *Annals of Tourism Research, 6*(4), 408–424.

Dann, G. M. S. (1977). Anomie, ego-enhancement and tourism. *Annals of Tourism Research, 4*, 184–194.

Decrop, A. (1999). Tourists' decision-making and behavior processes. In A. Pizam & Y. Mansfeld (Eds.), *Consumer Behavior in Travel and Tourism* (pp. 103–133). Haworth Press.

Dellaert, B. G. C., Arentze, T. A., & Horeni, O. (2014). Tourists' mental representations of complex travel decision problems. *Journal of Travel Research, 53*(1), 3–11.

Dellaert, B. G. C., Ettema, D. F., & Lindh, C. (1998). Multi-faceted tourist travel decisions: A constraint-based conceptual framework to describe tourists' sequential choices of travel components. *Tourism Management, 19*(4), 13–32.

Edwards, W. (1954). The theory of decision making. *Psychological Bulletin, 51*(4), 380.

Engel, J., Kollat, D., & Blackwell, R. (1978). *Consumer behaviour*. Dryden.

Fesenmaier, D. R., & Jeng, J. (2000). Assessing structure in the pleasure trip planning process. *Tourism Analysis, 5*, 13–27.

Goossens, C. (2000). Tourism information and pleasure motivation. *Annals of Tourism Research, 27*(2), 301–321.

Hill, T. H. (2000). *Investigating cognitive distance and long-haul destination*. http//www. Tourism wu-wien.ac.at

Horner, S., & Swarbrooke, J. (1996). *Marketing tourism, hospitality, and leisure in Europe*. International Thomson Business Press.

Hsu, T. K., Tsai, Y. F., & Wu, H. H. (2009). The preference analysis for tourist choice of destination: A case study of Taiwan. *Tourism Management, 30*(2), 288–297.

Hyde, K. F. (2004). A duality in vacation decision-making. In G. I. Crouch, R. R. Perdue, H. J. P. Timmermans, & M. Uysal (Eds.), *Consumer psychology of tourism, hospitality and leisure* (pp. 161–168). CABI.

Iso-Ahola, S. E. (1982). Towards a social psychology of tourist motivation: A rejoinder. *Annals of Tourism Research, 9*, 256–261.

Jafari, J. (2013). *Encyclopedia of tourism*. Routledge.

Jang, S., Bai, B., Hu, C., & Wu, C.-M. E. (2009). Affect, travel motivation, and travel intention: A senior market. *Journal of Hospitality & Tourism Research, 33*(1), 51–73.

Jang, S., & Cai, L. A. (2002). Travel motivations and destination choice: A study of British outbound market. *Journal of Travel & Tourism Marketing, 13*(3), 111–133.

Jeng, J., & Fesenmaier, D. R. (2002). Conceptualizing the travel decision-making hierarchy: A review of recent developments. *Tourism Analysis, 7*(1), 15–32.

Khan, M. J., Chelliah, S., & Ahmed, S. (2017). Factors influencing destination image and visit intention among young women travellers: Role of travel

motivation, perceived risks, and travel constraints. *Asia Pacific Journal of Tourism Research, 22*(11), 1139–1155.
Kotler, P., Bowen, J., & Makens, J. (2006). *Marketing for hospitality and tourism* (4th ed.). Pearson Education International.
Mansfeld, Y. (1992). From motivation to actual travel. *Annals of Tourism Research, 19*, 38–58.
March, J. G., & Simon, H. A. (1958). *Organizations*. Wiley.
Martin, D., & Woodside, A. G. (2012). Structure and process modeling of seemingly unstructured leisure-travel decisions and behavior. *International Journal of Contemporary Hospitality Management, 24*(6), 855–872.
Maslow, A. (1954). *Motivation and personality*. Harper & Row.
McNamara, K. E., & Prideaux, B. (2010). A typology of solo independent women travellers. *International Journal of Tourism Research, 12*(3), 253–264.
Moutinho, L. (1987). Consumer behavior in tourism. *European Journal of Marketing, 21*(10), 5–44.
Mutanga, C. N., Vengesayi, S., Chikuta, O., Muboko, N., & Gandiwa, E. (2017). Travel motivation and tourist satisfaction with wildlife tourism experiences in Gonarezhou and Matusadona National Parks, Zimbabwe. *Journal of Outdoor Recreation and Tourism, 20*, 1–18.
Nassar, M. A., Mostafa, M. M., & Reisinger, Y. (2015). Factors influencing travel to Islamic destinations: An empirical analysis of Kuwaiti nationals. *International Journal of Culture, Tourism and Hospitality Research, 9*(1), 36–53.
Nejati, M., & Mohamed, B. (2014). Investigating the key factors influencing the travel decisions of international tourists. *International Journal of Leisure and Tourism Marketing, 4*(2), 106–116.
Osei, B. A., & Abenyin, A. N. (2016). Applying the Engell–Kollat–Blackwell model in understanding international tourists' use of social media for travel decision to Ghana. *Information Technology & Tourism, 16*(3), 265–284.
Osei, B. A., Mensah, I., & Amenumey, E. K. (2018). Utilisation of social media by international tourists to Ghana. *Anatolia, 29*(3), 411–421.
Papatheodorou, A. (2006). *Managing tourism destinations*. Edward Elgar.
Park, J., Musa, G., Moghavvemi, S., Thirumoorthi, T., Taha, A. Z., Mohtar, M., & Sarker, M. M. (2019). Travel motivation among cross border tourists: Case study of Langkawi. *Tourism Management Perspectives, 31*, 63–71.
Pearce, P. (1988). *The Ulysses factor: Evaluating visitors in tourist settings*. Springer-Verlag.

Rathonyi, G. (2013). Influence of social media on tourism—Especially among students of the University of Debrecen. *Applied Studies in Agribusiness and Commerce, 10*, 105–112.
Schmoll, G. A. (1977). *Tourism promotion*. Tourism International Press.
Segumpan, R. G., Abu Zahari, J. S., & Jamaluddin, M. M. (2010). Tourism among families in Northern Peninsular Malaysia. *Asia-Pacific Social Science Review, 8*(2), 129–139.
Sellick, M. C. (2004). Discovery, connection, nostalgia: Key travel motives within the senior market. *Journal of Travel & Tourism Marketing, 17*, 55–71.
Seyidov, J., & Adomaitienė, R. (2017). Factors influencing local tourists' decision-making on choosing a destination: A case of Azerbaijan. *Ekonomika, 95*(3), 112–127.
Simon, H. A. (1955). A behavioral model of rational choice. *The Quarterly Journal of Economics, 69*(1), 99–118.
Sirakaya, E., & Woodside, A. G. (2005). Building and testing theories of decision making by travellers. *Tourism Management, 26*(6), 815–832.
Um, S., & Crompton, J. L. (1990). Attitude determinants in tourism destination choice. *Annals of Tourism Research, 17*, 432–448.
Van Raaij, W. F., & Francken, A. A. (1984). Vacation decisions, activities, and satisfactions. *Annals of Tourism Research, 11*, 101–112.
Van Vuuren, C., & Slabbert, E. (2011). Travel motivations and behaviour of tourists to a South African resort. In *International Conference on Tourism & Management Studies* (pp. 295–304).
Von Neumann, J., & Morgenstern, O. (1944). *Theory of games and economic behaviour*. Princeton University Press.
Woodside, A. G., & Lysonski, S. (1989). A general model of traveller destination choice. *Journal of Travel Research, 27*(4), 8–14.
Yoo, C.-K., Yoon, D., & Park, E. (2018). Tourist motivation: An integral approach to destination choices. *Tourism Review, 73*(2), 169–185.

6

Service Quality, Tourist Satisfaction, and Destination Loyalty in Emerging Economies

Eunice Fay Amissah, Evelyn Addison-Akotoye, and Sarah Blankson-Stiles-Ocran

Introduction

Tourism has been proven to drive economic growth and development, act as a force for good in the world, support United Nations 2030 Agenda for Sustainable Development, and contribute to the achievement of the Sustainable Development Goals (SDGs). Tourism has played a vital role in the development of industrialised countries. Tourism is now a huge

E. F. Amissah • E. Addison-Akotoye (✉)
Department of Hospitality and Tourism Management, University of Cape Coast, Cape Coast, Ghana
e-mail: eamissah@ucc.edu.gh; evelyn.addison@ucc.edu.gh

S. Blankson-Stiles-Ocran
Department of Hospitality and Tourism Management, Ho Technical University, Ho, Ghana
e-mail: sblankson-stiles-ocran@htu.edu.gh

I. Mensah et al. (eds.), *Marketing Tourist Destinations in Emerging Economies*, Palgrave Studies of Marketing in Emerging Economies,
https://doi.org/10.1007/978-3-030-83711-2_6

contributor to foreign exchange earnings and jobs to emerging economies. For instance, tourism is regarded as one of the most important industries and a key economic driver of many emerging tourist destinations such as Hong Kong, India, Ghana, South Africa, Kenya, Tanzania, Turkey, and Egypt. Emerging economies should, therefore, aim at cultivating loyalty intentions among tourists, which become a primary lever for steering their distinguished tourist products and services competitively. According to Campón-Cerro et al. (2017, p. 1), "Tourism destinations now must operate in extremely competitive markets, forcing destination managers to seek out innovative strategies and sustainable competitive advantage, including cultivating tourist loyalty". Hence, this chapter explores issues related to destination quality, tourist satisfaction, and destination loyalty in emerging tourist destinations. Specifically, the chapter begins with a discussion of the prospects and challenges for emerging destinations in delivering service quality. This is followed by an explication of tourism's contribution to global and emerging economies. The chapter further elucidates the concepts of service, destination quality, perceived value, satisfaction, and loyalty, as well as the linkages among them. Observations of the influence of destination quality, perceived value, and tourists' satisfaction on tourists' loyalty to destinations in some selected emerging tourist destinations are also presented. The chapter concludes by examining the implications and strategies for enhancing service quality and destination loyalty in emerging economies.

Prospects of Tourism and Challenges of Service Quality in Emerging Economies

The tourism sector is wide and comprises thousands of large and small industries, companies, and employers ranging from multimillion-dollar corporations to small individually owned businesses. The industry's continued growth and diversification, coupled with increasing globalisation, a relatively strong global economy, growing middle classes and rapid urbanisation in emerging economies, affordable travel and visa facilitation, as well as technological advances and new business models, contributed to it becoming one of the largest and fastest-growing economic sectors in the world (The World Tourism Organization [UNWTO], 2021). According to the World Travel and Tourism Council (WTTC; 2020), in 2019, the industry grew by 4%, was the second fastest-growing sector after manufacturing, and contributed 10.3% (US$8.9 trillion) to the global GDP while creating 330 million direct and indirect jobs.

Most developing and emerging economies are fast-growing in terms of the contribution of tourism to GDP. According to WTTC (2020), the sector accounted for US$ 1.7 trillion visitor exports (6.8% of total global exports, 28.3% of total global service exports) and US$948 billion capital investment (4.3% of total investments) in 2019. Tourism growth was deemed to be strongest in emerging economies (+4.4%), with tourist arrivals foreseen to increase at twice the rate of those in advanced economies (+2.2%) between 2010 and 2030 (UNWTO, 2011). Expectedly, international tourism was to increase in Asia and the Pacific (to 30% up from 22% in 2010), the Middle East (to 8% up from 6%), and Africa (to 7% up from 5%) (UNWTO, 2011). In 2017, the market share of international tourism in emerging economies was again projected to reach 57%, upon reaching 45% in 2016 (up from 30% in 1980) (UNWTO, 2017). However, with the emergence of the Covid-19 pandemic, Rajan Raghuram, a chief economist with IMF and advisor to the Indian government, has specified that the GDP growth of emerging economies is expected to reduce from an average of 5.5% in 2000-2019 to around 3.5% in 2020-2040. He added that growth will still be faster than that in the developed world, but incomes of emerging economies will converge more slowly than previously (Kohlmann, 2020).

Despite the enormous contribution of the tourism sector to emerging economies, most of them have had to surmount many compelling challenges to remain attractive and competitive and cultivate loyalty among tourists. Notable among these challenges are political unrest, susceptibility to terrorist attacks, economic recession, high levels of crime, lack of product diversity, fragmented government and regulatory infrastructure, youthful population, expanding workforce, and rapid urbanisation. These challenges are, however, more evident in the new emerging destinations. These challenges usually heighten stereotypes, prejudices, and negative destination image, which hinder effective destination marketing, resulting in fluctuations in tourist arrivals and revenue. Recognising the benefits derived from international tourism, some new emerging economies of Africa, for instance, attempted to solve their challenges by strengthening the capacity of their tourism ministries and authorities to reform the tourism industry by leveraging the country's key tourism resources. For example, improved security in the years following 2015 and intensified marketing efforts by Kenya Tourist Board afforded the tourism sector a ripe opportunity to rebound with 3.9% growth in arrivals (Government of Kenya [GoK], 2019).

Looking beyond the negative impacts of the aforementioned challenges, there are still prospects for tourism development. From the supply perspective, the tourism industry is now characterised by intense rivalry, with most players offering the same or similar products. On the other hand, the industry has become customer driven, with different tourists demanding better services that commensurate their sacrifices, lifestyle, purpose of travel, and family size. The shift in the supply and demand perspectives offers some prospects, as well as poses new challenges of service quality to all stakeholders. The factors that will support the continuity and sustainability of tourism in emerging economies are unique attractions, intense regional and international competition, youthful population, and private sector involvement. Inherent within these prospects are issues of service quality that need to be addressed to achieve tourist satisfaction and loyalty.

First, there is empirical evidence that product or service uniqueness predicts consumer purchase intentions (Rubera et al., 2011). This trigger is leveraged by most businesses to distinguish their products and services from their competitors. In the case of tourism, attractions form a key component of the supply side, which pulls most tourists to such destinations. Most emerging destinations have unique attributes, which differentiate them from mature destinations (Esu & Ebitu, 2010). These unique characteristics provide novel experiences to tourists from major source markets, mostly in the experienced economies. Their uniqueness lies mostly in the natural and cultural resources, and hospitality. Examples include ancient temples, forts and castles, caves, famous national reserves and parks, pristine beaches, festivals and good climate. In addition to the natural and cultural assets, some emerging destinations also offer unique experiences in shopping, events, and gastronomy. For instance, while the name Kenya resonates with the word 'Safari', Hong Kong is considered the hub of international convention and exhibition. Ghana is also known for heritage tourism, with a focus on the history of the slave route; Egypt is famous for the pyramids, temples, and tombs; India is renowned for its sacred sites, cultural and historical treasures; and Turkey is known for its historical sites and seaside resorts. A combination of these unique man-made and natural attractions makes emerging destinations very attractive and ripe for investments. Critical to this discussion are issues of accessibility, supporting amenities, and other activities within a destination. Tourists visiting a unique site must have good access to the destinations and attractions. Also, they will require quality accommodation and diverse activities to drive their spending and length of stay within the destination.

Second, there is growing regional and international competition among destinations entering the tourist market. Tourists will definitely trade one destination for another, given their purpose of travel, budget and perceived value judgement of the competing destinations, among other reasons. Competition is necessitated by two main conditions: (a) the number of tourist destinations available and (b) destinations offering similar attractions. Foremost, many destinations mean a wider range of destination options available to tourists, which introduces elements of choice and preferences, irrespective of the similarity or dissimilarity of the attractions. The next level of competition sets in when tourists have to choose between two destinations offering similar experiences. The examples from the preceding paragraph point to the fact that the attractions, albeit unique in terms of experience, are not exclusive to destinations. For example, Ghana, India, and Turkey offer historical attractions and experiences, which leaves the tourist with a choice. Similarly, within the African market alone, more than ten countries promise unique safari experiences to tourists. There is empirical evidence that competition is one of the key factors driving service quality in the global tourism market (Burch et al., 1995). Hence, to attract and retain a good market share, destinations must take a keen interest in enhancing service quality.

Emerging destinations are also characterised by a youthful population, which is good for the human resource potentials for a human-centred industry like tourism. This means that there would not be shortages in terms of the workforce in the sector, especially when these countries have institutions running tourism and hospitality management, and hotel and catering management programmes. However, in most new emerging economies, the issue of unprofessional and unskilled workforce has been a major problem facing the industry, with an ensuing negative effect of poor service quality. To solve this problem, most new emerging economies have fallen on expatriates to fill the jobs and hopefully train the locals to take over from them.

Finally, the involvement of the private sector represents another prospect for tourism development in emerging economies. Their contribution is vital for the growth of tourism and hospitality because of the profit interest underlying their business operations. This interest drives them to assiduously work with the awareness of clients' needs and wants for their business survival. While this is good for the tourism industry, it poses an inherent service quality challenge, especially in new emerging destinations. For example, the hospitality sub-sector of the tourism industry in Ghana is characterised by small- and medium-sized businesses, mostly lacking the capacity to attract and access skilled labour. This renders the industry very vulnerable to poor service quality and customer dissatisfaction.

Concepts

Service Quality

Service quality, as a concept, is one of the critical issues in the tourism industry, as it poses one of the major challenges faced by the management of tourism businesses (Mensah & Dei Mensah, 2013). With an increase in demand for tourism activities, tourism businesses have become highly competitive. This presupposes that to survive in the industry, tourism businesses have to meet guest expectations through the delivery of quality services. As such, service quality, as a concept, is still a very significant research topic in the tourism and hospitality sector.

Earlier proponents of the concept adopted the service gap approach to define service quality as a function of the difference between customers' expectation of service and their perception about the actual service experienced (Parasuraman et al., 1988). They considered the concept a fundamental factor to the success of any service-based business. This concept is very relevant to the tourism and hospitality industry because it is mainly service based. Underlying this concept is the notion that service quality is judged by the receivers of the service experience. Thus, it starts with the customer's expectations of any specific service and ends with their post-consumption perception. The subject of service quality is wide and varied (Narteh, 2018) and has been operationalised from different perspectives. Gundersen et al. (1996) argued that a customer's experience is a product of both the tangible and intangible aspects, whereas Grönroos (1994) distinguished between technical quality and functional quality as components of the service image delivery. Technical quality is what the consumer actually receives from the service provider, such as the quality of the food served in a restaurant, while functional quality, on the other hand, is how the technical component of the service is delivered to the consumer.

Kotler and Keller (2013) identified a third component of quality—societal (ethical) quality, which is described as credence quality. This type of quality does not lend itself to advance evaluation prior to the actual purchase. In this light, emerging tourist destinations concerned about quality service must combine, balance, and juggle the tangible and intangible products to deliver quality service experiences to tourists. This must be approached by taking cognisance of tourists' emotional and hedonic predispositions regarding their entire experience, including their interface with physical surroundings, service providers, other customers, and other participants in the tourist service (Chang et al., 2016). Consequently, delivering service quality can promote tourist advocacy and foster destination loyalty.

Destination Quality

Destination quality, like service quality, is measured using tourist expectations and actual service performance. Destination quality is broader and covers the quality of service received from all aspects of a destination, including transport, food, accommodation, attractions and tour guiding. In this regard, Buhalis (2000) identified six components for analysing tourism destinations. These are (a) attractions (a primary offer of tourism that due to its amount, quality and attractiveness activate attendance, for instance, natural, man-made, and cultural-historical potential); (b) accessibility (a general infrastructure which enables access to the destination, travelling to the attractions in the destination such as transportation routes, terminals and vehicles); (c) amenities (superstructure and infrastructure of tourism that enable the stay in the destination and utilising its attractions, e.g. accommodation and catering facilities, sports, recreational, cultural-social facilities); (d) available packages (prepared products and product packages by intermediaries); (e) activities (all activities available at the destination and what consumers engage in during their visit); and (f) ancillary services (services used mainly by local inhabitants and tourists, such as telecommunication, medical and banking services). The characteristic components of destinations identified by Buhalis stress the fact that an evaluation of destination service quality should balance issues of both technical quality and functional quality components.

Destination quality is critical to achieving positive word of mouth and loyalty to a destination. Studies conducted in emerging economies suggest a positive link between destination quality and destination loyalty such that if a tourist perceives a destination like Kenya, Turkey, and Ghana to be of quality, then they are more likely to recommend the destination to other potential tourists and revisit the destination (Nguthi et al., 2021).

Perceived Value

Perceived value is the sacrifice made by the consumer to obtain a product or service (Mohamad et al., 2019). It is more about the trade-off between perceived benefits and perceived costs (Lovelock, 2000). There are two approaches to understanding the concept of perceived value: unidimensional and multidimensional (Hallak et al., 2018). The unidimensional idea is based on the neoclassical economic theory and assumes that consumers are rational beings who are governed by maximising their utility when it comes to choosing between different products/services (Chiu et al., 2005). Under this utilitarian perspective, consumers' perceived value of a product and service is generally determined by the difference between performance (benefits consumers receive from the product/service) and the sacrifices they make to acquire that product/service. Sacrifices involve the monetary and non-monetary cost (such as time and effort expended) incurred by the customer to acquire the product (Mohamad et al., 2019). The multidimensional idea encompasses both the cognitive and affective facets of a product and service.

Tourist Satisfaction

Tourist satisfaction, just like service quality, plays a key role in the success and sustainability of tourist destinations. Satisfaction is defined as the extent to which a service "provides a pleasurable level of consumption-related fulfilment" (Oliver, 1997, p. 13) and has been used as a performance indicator of a destination's product and services. Satisfactions have been evaluated from four main approaches: (a) equity theory, (b) norm theory, (c) the perceived overall performance theory, and (d) the disconfirmation paradigm. Among these four, the expectation-disconfirmation paradigm (Oliver, 2006) has been widely used in most satisfaction studies involving four dimensions of satisfaction: (a) expectation, (b) performance, (c) disconfirmation, and (d) satisfaction. This paradigm is premised on the disparity between the prior expectations and the actual performance. Accordingly, satisfaction occurs when post-purchase experiences exceed prepurchase expectations (i.e. positive disconfirmation). On the other hand, dissatisfaction occurs when performance falls short of prior expectations (i.e. negative disconfirmation).

Tourist satisfaction is perceived as a tourist's overall emotional response to the whole destination experience following the visit (Kim & Thapa, 2018). However, it suffices to note that satisfaction is not inherent in the individual or the product but is a socially constructed response to the relationship between a customer, the product, and the service provider. Tourist satisfaction has been noted as a key consideration for destination marketing, as it achieves three things: it (a) influences tourists' choice of a destination, (b) convinces them to consume more similar kinds of products and (c) motivates them to return to the places visited earlier. However, tourist satisfaction has become the key challenge for destination managers since the marketing strategy is focused on attracting more tourists to the destination (Meleddu et al., 2015).

Destination Loyalty

Loyalty is the prospect that consumers will use a certain type of brand in the coming period, regardless of the market opportunities and efforts of competitors offering the same or similar products and services (Veljković & Đorđević, 2010). Loyalty is perceived as desirable from a destination perspective because it has often been considered an important factor for a destination's competitiveness. It is also a good indicator of tourists' response to the destination's product. In other words, tourists would show a positive attitude towards a destination as a sign of their loyalty which is usually expressed through positive word of mouth. The powerful effect of word of mouth is highly critical for destination competitiveness owing to the fact that most travellers consider such travel experience reviews from friends and relatives to be very reliable. According to Wang et al. (2011), destination loyalty is defined as recommending the destination to others by a satisfied tourist. Loyal tourists continually act as influential agents for promoting a destination, which represents their highest level of commitment towards a destination. Hence, building long-term relationships with tourists has become an indispensable part of effective destination management.

Approaches to Destination Loyalty

There are three main approaches mostly used in the conceptualisation of destination loyalty: (a) attitudinal, (b) behavioural, and (c) composite.

Attitudinal

This approach considers the willingness to spread positive word of mouth and intention to recommend the destination to friends and family. The intention to recommend refers to "the informal person-to-person communication between a perceived non-commercial communicator and a receiver regarding a brand, a product, an organisation, or a service" (Harrison-Walker, 2001, p. 63). This intention to recommend the destination is of great relevance because it reduces the perceived risk involved in the decision of choosing a tourist destination, which brings about a strong need for reliable information sources.

Behavioural

Destination loyalty is commonly assessed based on tourists' behavioural intention, which is measured in terms of tourists' intention to revisit the destination (Gao & Lai, 2015). In the marketing and tourism literature, tourist loyalty is fully interchangeable with behavioural intentions. Given the current competition among emerging tourist destinations locally and globally to capture a significant proportion of the tourist market share (Sangpikul, 2018), destinations are resorting to providing quality and valuable experiences, tourist satisfaction, and destination loyalty as a competitive advantage. To a large extent, the success of a tourist destination depends on tourists making repeat visits to that destination (Alrawadieh et al., 2019).

Composite

The approach integrates both behavioural and attitudinal dimensions of loyalty. By extension, loyalty can be interpreted by tourist's response to both attitudinal and behavioural triggers. Using this approach, a tourist who visits and has loyalty towards a particular destination must exhibit a positive attitude towards the destination by the intention to revisit, willingness to spread positive word of mouth, and intention to recommend the destination to family and friends.

Importance and Benefits of Loyalty

Some researchers have argued in the past that the issue of loyalty is of major concern to countries that have reached the maturity period in the tourism destination life cycle probably because, at that point, most tourists would no longer find such destinations attractive and would turn to discover new ones. If this argument still holds, we could easily conclude that emerging tourism economies could relax their marketing efforts because they stand a better chance of attracting newcomers to the tourist market. However, such an argument could be erroneous in the sense that the action taken by tourists after experiencing a new destination could lead the destination on two main paths—to be attractive or otherwise to prospective tourists. Thus, a customer (satisfied or not) is likely to tell his/her story to others. If the tourist is satisfied, the chances are that he or she will recommend the destination. The reverse holds true for unsatisfied tourists.

In line with this reasoning, there is an opportunity to attract more tourists if the destination builds loyalty among tourists. Thus, rather than committing more resources to attract first-time visitors, it is worthwhile to retain repeat visitors. After all, studies have shown that repeat visitors have a higher propensity to revisit and recommend the destination to others as compared to first-time visitors (Chi, 2012). It is a known fact that a reduction in tourist defection will ensure a constant flow of inbound travellers and generate a significant increase in tourist revenues (Hallak et al., 2018). Loyal tourists pay less attention to competing destination, which has significant bearings on both current and future values for the competitiveness and sustainability of emerging destinations. Nilplub et al. (2016) highlight four main benefits of building loyalty among travellers:

1. The destination spends less to attract repeat travellers than to woo newcomers.
2. Repeat travellers spend more than the new counterparts at the same destination.

3. The likelihood of loyal tourists repeating a visit to the same destination is higher than that among newcomers.
4. Repeat visitors would most likely recommend the destination to others.

Service Quality, Perceived Value, Tourist Satisfaction, and Destination Loyalty

It is envisaged that competition among destinations will become even greater in the years to come (Wu & Li, 2015). Thus, having a better understanding of why tourists become tied to a destination and what drives loyalty is crucial. As posited by Hallak et al. (2018), tourists' perceptions of service quality and value of a destination affect satisfaction, generate customer referrals, and repeat visits. This does not only affect the destination but also impacts the long-term performance of tourism-related enterprises (Al-Msallam, 2020). Hence, understanding the relationships among service quality, perceived value, tourist satisfaction, and destination loyalty has practical implications for the effective marketing of emerging destinations. Figure 6.1 depicts these established relationships.

Service quality is positively related to perceived value, and significantly impacts satisfaction and loyalty, as empirically evidenced by the work of Chi et al. (2020). In most studies, however, satisfaction has been found to either have a direct relationship or mediate the relationship between service quality and loyalty. Perić et al. (2020) reported a direct correlation between service quality and satisfaction, as well as between satisfaction and loyalty. Similarly, Wang et al. (2017) found a direct effect of satisfaction on loyalty. On the other hand, El-Adly and Eid (2016) found an indirect influence, and, thus, customer satisfaction played a mediating role between customer's perceived value and loyalty. Nilplub et al. (2016) also revealed that the satisfaction of tourists fully mediates the effects of push and pull motivations and perceived value for money on destination loyalty, and that it partially mediates the relationship between perceived service quality and destination loyalty. Further, it is quite unlikely that unsatisfied tourists will

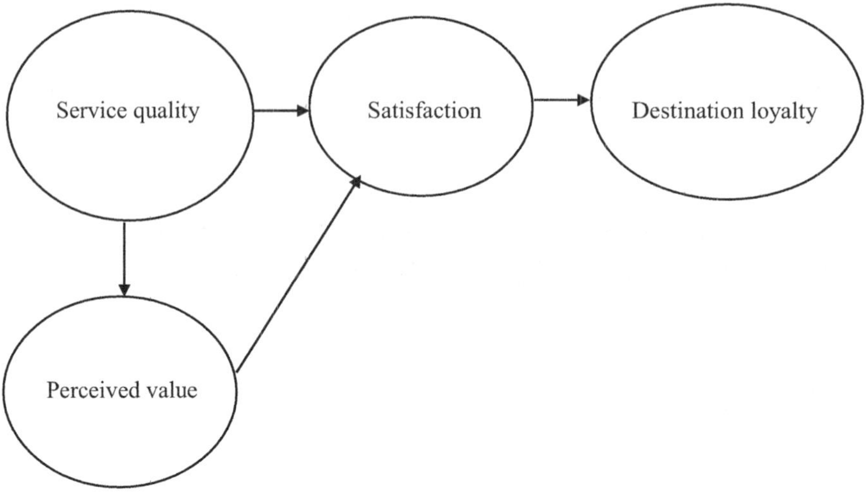

Fig. 6.1 Proposed framework of linkages between service quality, perceived value, tourist satisfaction, and destination loyalty

return to destinations or recommend that others visit them (Nilplub et al., 2016), but it has been established that increasing travel satisfaction leads to increasing tourists' destination loyalty (Kim & Thapa, 2018).

Drivers of Destination Loyalty in Emerging Economies

It cannot be overemphasised that cultivating loyalty in the tourism industry is an important factor for destination competitiveness and sustainability. It is worth noting that different categories of tourists are connected in different ways to particular destinations, which presupposes a complex connection. However, given that destinations offer differentiated products and services which produce different experiences for tourists and subsequently their loyalty intentions, it is necessary to understand how the antecedents interact to produce loyalty with reference to some selected emerging tourist destinations. Among the antecedents mostly studied, including destination attributes, perceived value, and perceived service quality, satisfaction has consistently emerged as the variable with the greatest direct impact on loyalty. Tourist destinations have remained competitive by exploiting the relevant features of the 'six A's' destination attribute classifications (attractions, accessibility, amenities, available packages, activities, and ancillary services) emphasised by Buhalis (2000).

With increasing pressure on emerging economies to create effective marketing strategies for tourism, it becomes necessary to first understand tourist perceived service quality and value, which can be linked directly to their derived satisfaction. Perceived service quality and value, and satisfaction have been treated as independent or mediating variables in most loyalty studies. Raza et al. (2020) assert that adding value to the service at a competitive price is a potent source of competitive advantage. Customers perceive a higher value in the service when they perceive the quality of service as greatly exceeding the costs they sacrificed to obtain the service. Su et al.'s (2021) study on authenticity, perceived value, and loyalty in marine tourism destinations in Zhoushan, Zhejiang Province, China, also reveals that destination service quality positively influences perceived value. Destinations can augment the perceived value of a product or service by either delivering a better service or reducing customer perceptions of the costs associated with using the service.

Regarding 'attractions', destinations endowed with natural attractions in emerging tourist destinations have cultivated destination loyalty by focusing on relevant attributes of the main attraction. For example, Bangladesh thrives as having one of the world's longest beaches by highlighting the unique beauty of their beaches, nearby islands, site tour attractions, and the reputation of destination quality. Kamrul and Shamsul (2019) revealed that tourists' perceived service quality and perceived value had significant positive effects on their revisit intentions. Specifically, tourists described beach destinations in Bangladesh as exciting, desirable, and pleasant, so they will recommend and visit the destination again. In addition, perceived value showed a more positive effect than perceived service quality on tourist satisfaction with visiting beach destinations. Likewise, tourist satisfaction had a significant influence on tourist loyalty.

Similarly, Gaikwad et al. (2020) studied tourists' satisfaction and destination loyalty at the Ellora Caves, one of the largest rock-cut monastery cave temples in the world and a UNESCO world heritage site in Aurangabad district of Maharashtra, India. They alluded that tourists were overwhelmed by the highly appealing monument, scenic beauty around Ellora Caves, architectural beauty, and overall tourist satisfaction at Ellora Caves. However, the lowest level of satisfaction was observed for attributes such as shopping, connectivity, public utilities, availability of tourist information centre, promptness at the ticket window, availability of recreation, and entertainment facilities. Further, harassment of the tourists by local vendors to sell their products forcefully and the sale of fake semi-precious stones in the name of original stones were observed to negatively affect tourists' overall perception of the destination. Nonetheless, the study revealed a positive and strong relationship between overall satisfaction and destination loyalty.

Emerging destinations that offer mostly eco-tourism products rely heavily on tour guides to facilitate tours while delivering the promises made by the DMOs. These tour guides have longer interactions with the tourists. Hence, they form an essential part of the tourism product and play ambassadorial roles contributing to the destination quality, tourist satisfaction, and subsequently loyalty. The significance of tour guides in enhancing a destination's image has been widely recognised in the literature (Arasli & Baradarani, 2014). A study conducted by Nguya et al. (2021) on the influence of tour guide service quality on customer satisfaction in Narok County in Kenya revealed that punctuality and a keen sense of direction were the most important competencies of tour guides influencing tourist satisfaction. In the same study, it was reported that the delivery of promised service ranked highest among other core service delivery variables, including honesty and trustworthiness, ability to solve problems, and knowledge of the destination. Finally, the study revealed from the SERVQUAL gap analysis that tourists were dissatisfied with the tour guiding services provided.

Similarly, Ariya et al. (2020) evaluated tourist guides' satisfaction with wildlife tourism destinations' attributes and revisit intentions at the Lake Nakuru National Park in Kenya. The study revealed a positive but weak relationship between satisfaction and revisit intention, indicating that satisfaction was not a good indicator for revisit intentions. Specifically, they found that tourist guides were mostly satisfied with the wildlife resource attributes such as uniqueness, variety, and abundance. This was closely followed by the cost of attraction, park image, and accessibility dimension. Ariya et al. alluded to the fact that within the study setting, tourist guides sometimes do not have access to tourists and visitors to influence their revisit intentions. This is because tour operators usually prefer the services of private driver guides for tourists and visitors who come on packaged tours and with a predetermined itinerary.

Also, satisfaction with 'activities' at a destination influences destination loyalty. A study concluded that tourists who mostly enjoyed all the main tourist activities in KwaZulu-Natal in South Africa were found to be satisfied (Ezeuduji & Mhlongo, 2019). Consequently, tourists who are satisfied with a destination are more likely to return to that destination, become loyal, and recommend a destination to other people such as their friends and family members. For example, the topmost loyalty statements made by tourists are the following: "I would recommend KwaZulu-Natal to my friends and family", "I would consider visiting KwaZulu-Natal in the future", and "I am in love with KwaZulu-Natal". These results indicate a high destination loyalty from the responding tourists.

Additionally, international business events have been identified as one of the opportunities to enhance destination competitiveness (Weru & Njoroge, 2021). A well-organised and executed event has the potential of increasing a host destination's image while wearing away the existing stereotype (Dragin-Jensen & Kwiatkowski, 2019). A study conducted by Weru and Njoroge (2021) in Kenya identified event venue facilities, destination accessibility, safety and security, attractions, affordability of services, and accommodation facilities as the dimensions of business event experience, which influence the host destination image. It was, thus, concluded that business event experiences influenced tourists' perceived value and overall image of the destination. Moreover, Hong Kong has harnessed the potential of international business events to its advantage and was considered the hub of international convention and exhibition. Hong Kong plays host to a number of international conventions and exhibitions owing to its wide selections of venues and experience in managing large meetings (Tang et al., 2011).

Furthermore, while most tourists have been found to express a higher level of satisfaction with attractions, 'Access, Amenities and Ancillary services' have induced negative reactions. Bhat and Darzi (2018) conducted a study in Jammu, Kashmir, and Ladakh in India and found that tourists were satisfied with the destination attributes like scenic beauty, architectural beauty, cave's visitor orientation centre, and conservation of the monument. However, tourists were dissatisfied with infrastructural attributes like connectivity, public convenience, transport operators, budget hotels, public utilities, recreation and entertainment facilities, and bank and ATM facilities.

Likewise, a study by Yeshi (2019) measuring the level of tourist satisfaction with Bhutan as a destination showed that road and travel conditions were perceived by visitors to be in bad shape that required urgent attention of the government to give a facelift to national highways. The study also reported that while tour guides were doing well in all areas of duties, the accommodation facilities scored badly. Though tourists were satisfied with the culture and beauty of the natural environment, they expected varied tourism products and services. Howbeit, tourists were more willing to revisit and recommend the destination.

Studies in Ghana have come out with diverse findings on tourist satisfaction with certain destination attributes. For instance, empathy factors like employees understanding the specific needs of guests and delivering personalised attention (Amissah, 2013), accommodation, community interaction, and souvenir shopping (Preko et al., 2020) were significant to explain international tourist satisfaction of accommodation in Cape Coast. Likewise, while Adongo et al.'s (2015) study on international tourists' experience of Ghanaian cuisine revealed that local food has a positive and significant influence on willingness to recommend, Preko et al. (2020) found that food and tour services had a negative effect on tourists' satisfaction and loyalty.

Conclusions and Recommendations

The chapter has drawn evidence from previous studies to argue that service quality, satisfaction, and loyalty are not issues for only old destinations. This chapter further provides insights into the antecedents of customer loyalty, that is, service quality, perceived value, and customer satisfaction, and more specifically, the relationships between antecedents and customer loyalty. The benefits of destination loyalty make it worthwhile for emerging economies to cultivate loyalty among tourists. The tourism industry, more than any other service industry, has the potential to elicit strong emotional and experiential reactions from tourists and travellers (Nilplub et al., 2016). This goes a long way to influence their satisfaction level and foster loyalty. With efforts by the various tourism stakeholders, particularly government, local residents, businesses and media, tourist destinations could enhance their attractiveness and competitiveness. The long-term competitiveness of destinations depends on the willingness of all stakeholders to minimise the service quality challenges and maximise the paybacks from the opportunities to reinvent service delivery. Innovative and competitive emerging tourist destinations must offer new service standards by looking beyond the traditional service delivery and marketing strategies. As clearly stated by Hikmah et al. (2018), in the context of tourism consumption, as the homogeneity of tourism resources increases, the uniqueness of tourism resources tends to have a significant bearing on tourists' loyalty.

From the analysis, service quality, destination quality, tourists' perceived value, satisfaction, and loyalty are significant aspects that require careful consideration when planning tourism development strategies. Accordingly, this leads to increased levels of perceived value, high levels of satisfaction, and enhanced customer loyalty, which is in turn reflected in increased sales, positive word of mouth, improved reputation, and destination loyalty. Largely, this creates opportunities for tourism businesses in emerging economies to benefit significantly from tourism and thrive. Thus, emerging destinations will be able to improve their financial position, achieve higher market shares, maintain their competitive advantage, attract new development resources and become sustainable on a long-term basis. Consequentially, emerging destinations need to understand the factors that determine the destination quality to be able to achieve customer satisfaction and reduce perceived risk through improved product and quality. These factors, which tourists perceive as important for quality, are critical.

It was observed that perceived quality (attractions, beaches, road and transport conditions, staff, food, accommodation, security, and tour services) would not affect destination loyalty directly but indirectly through the mediation of perceived value and customer satisfaction. Therefore, to enhance customers' perceived value and customer satisfaction, emerging destinations should start by protecting quality along the tourism value chain—tourists judge their experience with a destination taking into account the services received from all stakeholders along the tourism value chain in addition to the main attractions visited. It behoves on destinations to have a total quality management policy implemented by the government regulatory body mandated to carry out such responsibilities. Such policies must be implemented to the latter by sanctioning non-compliance and rewarding service excellence. Every stakeholder must show some level of commitment by aligning their quality standards (both tangibles and intangibles) to the national quality policy. This will help emerging destinations to gain competitive advantage by achieving customer satisfaction. In Ghana, for instance, the 2016 Legislative Instrument for Accommodation Enterprise (L.I. 2239) and Food and Beverage Enterprise (L.I. 2238) mandate the Ghana Tourism Authority Quality Assurance Department to be responsible for ensuring the service and product standards through inspections and performance monitoring.

It was also revealed that perceived value has an influence on tourist loyalty towards a destination. Specifically, tourists who hold high-value perceptions about a destination are more likely than others with low-value perceptions of the same destination to express loyalty. The findings indicate that higher service quality and greater value are of utmost importance in improving destination image in the context of emerging destinations. Improving visitors' perceptions of service quality and value with the view of enhancing their satisfaction is essential in retaining loyal visitors. Therefore, destination service providers must pay keen attention to key service indicators, especially hygiene and cleanness at beaches and nearby areas, employee-oriented services, and food and accommodation services, which can enhance the reputation of these destinations.

The chapter established that the service quality and value perceptions of tourists when visiting emerging destinations directly have an impact on increasing tourist satisfaction. This, thus, encourages tourist destination managers to continue to innovate services and present a variety of supporting attractions that can make tourists increasingly frequent the destinations. The perception of tourists on the quality of destinations, especially the supporting facilities provided by destination managers, fell below tourists' expectation. Destination managers have to improve the conditions of existing facilities and increase the provision of supporting facilities in destinations so that tourists feel comfortable and safe when visiting these destinations. Regarding safety and security challenges in some destinations, local community collaboration with the police, strong police presence and proper policing in these destinations will likely be effective in dealing with crime and reducing the negative perception of tourists towards such places. These improvements must be based on the needs and expectations of tourists expressed personally or through their post-experience reviews and feedbacks.

It is, therefore, recommended that destination management organisations and local businesses create an open information management system that allows easy integration of customer reviews and feedback for appropriate product and service improvement. Tourists who visit the destination must be seen as key actors in the creation and enjoyment of their experiences and hence can provide useful cues on their product and service expectations. Destinations must identify the gaps between the needs of tourists and the ability of the destination to meet such expectations. This can be done through periodic service audits and research, and avenues created by the destination to take guest complaints. At the micro-level, tourism-related businesses can obtain direct and regular customer feedback by asking customers to complete a service satisfaction survey. Information can also be solicited from online travel reservations and review platforms such as TripAdvisor or Expedia. At the macro-level, destinations can set up tourism information and call centres to provide standardised and verified tourism information as well as to receive enquiries and complaints about tourism encounters in the country. This will help tourism stakeholders to (a) anticipate the needs of tourists, (b) provide personalised tourism products/services, and (c) provide opportunities to respond quickly to customers' needs and requirements with new ideas and technology.

The analysis also revealed tourists' dissatisfaction with tour guiding services and promptness of services. To overcome such human factors in service quality, there is the need for specific training programmes and human resource development interventions to train a professional and skilled workforce. These interventions must aim at churning out a competent workforce who are not motivated by just short-term gains but long-term prospects. Training and development can be achieved through a private-public partnership. Training should be designed to meet specific service quality objectives and targeted at specific groups. Tourism regulators can engage private or public tourism and hospitality training institutions to facilitate free or subsidised certificate courses for tour guides, waiters, room attendants, drivers, and so on. Since most of these workforces are already engaged, on-the-job training and coaching could be the best forms of training opportunities in addition to the study leave opportunities given to workers to upgrade themselves. Training should not only target lower-level staff but management as well. Managers must be challenged to assess their management perspectives and create opportunities for employees to be creative and innovative.

References

Adongo, C. A., Anuga, S. W., & Dayour, F. (2015). Will they tell others to taste? International tourists' experience of Ghanaian cuisines. *Tourism Management Perspectives, 15*, 57–64.

Al-Msallam, S. (2020). The impact of tourists' emotions on satisfaction and destination loyalty—An integrative moderated mediation model: Tourists' experience in Switzerland. *Journal of Hospitality and Tourism Insights, 3*(5), 509–528. https://doi.org/10.1108/JHTI-11-2019-0126

Alrawadieh, Z., Alrawadieh, Z., & Kozak, M. (2019). Exploring the impact of tourist harassment on destination image, tourist expenditure, and destination loyalty. *Tourism Management, 73*, 13–20.

Amissah, E. (2013). Tourist satisfaction with hotel services in Cape Coast and Elmina, Ghana. *American Journal of Tourism Management, 2*(1), 26–33.

Arasli, H., & Baradarani, S. (2014). European tourist perspective on destination satisfaction in Jordan's industries. *Procedia-Social and Behavioral Sciences, 109*, 1416–1425.

Ariya, G., Sempele, C., & Wishitemi, B. (2020). An evaluation of wildlife tourism destinations' attributes, overall satisfaction and revisit Intentions: A tourist guides' perspective. *African Journal of Hospitality, Tourism and Leisure, 9*(6), 1220–1237. https://doi.org/10.46222/ajhtl.19770720-79

Bhat, S. A., & Darzi, M. A. (2018). Antecedents of tourist loyalty to tourist destinations: A mediated-moderation study. *International Journal of Tourism Cities, 4*(2), 261–278.

Buhalis, D. (2000). Marketing the competitive destination of the future. *Tourism Management, 21*(1), 97–116.

Burch, E., Rogers, H. P., & Underwood, J. (1995). Exploring SERVPERF: An empirical investigation of the importance-performance, service quality relationship in the uniform rental industry. http://saber.uca.edu/docs/proceedings11/95ama, 121.

Campón-Cerro, A. M., Hernández-Mogollón, J. M., & Alves, H. (2017). Sustainable improvement of competitiveness in rural tourism destinations: The quest for tourist loyalty in Spain. *Journal of Destination Marketing & Management, 6*(3), 252–266.

Chang, S., Gibson, H., & Sisson, L. (2016). *The role of satisfaction and involvement in loyalty: Comparing residents and tourists.* In Tourism travel and research association: Advancing Tourism Research Globally 2011 TTRA International Conference. Quebec City, Quebec, June 20–22.

Chi, C. G. Q. (2012). An examination of destination loyalty: Differences between first-time and repeat visitors. *Journal of Hospitality & Tourism Research, 36*(1), 3–24.

Chi, X., Lee, S. K., Ahn, Y. J., & Kiatkawsin, K. (2020). Tourist-perceived quality and loyalty intentions towards rural tourism in China. *Sustainability, 12*(9), 3614. https://doi.org/10.3390/su12093614

Chiu, C. M., Hsu, M. H., Sun, S. Y., Lin, T. C., & Sun, P. C. (2005). Usability, quality, value and e-learning continuance decisions. *Computers & Education, 45*, 399–416.

Dragin-Jensen, C., & Kwiatkowski, G. (2019). Image interplay between events and destinations. *Growth and Change, 50*(1), 446–469. https://doi.org/10.1111/grow.12272

El-Adly, M. I., & Eid, R. (2016). An empirical study of the relationship between shopping environment, customer perceived value, satisfaction, and loyalty in the UAE malls context. *Journal of Retailing and Consumer Services, 31*, 217–227.

Esu, B. B., & Ebitu, E. (2010). Promoting an emerging tourism destination. *Global Journal of Management and Business Research, 10*(1), 21–28.

Ezeuduji, I. O., & Mhlongo, P. S. (2019). Tourists' loyalty towards a tourism destination: The case of KwaZulu-Natal in South Africa. *Business Administration and Business Economics (Euro Economic), 2*(38), 272–281.

Gaikwad, S. T., Ragde, R., & Sawant, M. (2020). Tourist satisfaction and destination loyalty: A case study of the world heritage site of Ellora caves. *Atna Journal of Tourism Studies, 15*, 17–31. https://doi.org/10.12727/ajts.23.2

Gao, B., & Lai, I. K. W. (2015). The effects of transaction-specific satisfactions and integrated satisfaction on customer loyalty. *International Journal of Hospitality Management, 44*(1), 38–47.

Government of Kenya. (2019). Tourism sector performance report. http://ktb.go.ke/wp-content/uploads/2019/01/Tourism-Performance2018-Presentation-Final2.pdf

Grönroos, C. (1994). Quo vadis, marketing? Toward a relationship marketing paradigm. *Journal of Marketing Management, 10*, 347–360.

Gundersen, M. G., Heide, M., & Olson, U. H. (1996). Hotel guest satisfaction among business travelers: What are the important factors? *The Cornell Hotel and Restaurant Administration Quarterly, 37*(2), 72–81.

Hallak, R., Assaker, G., & El-Haddad, R. (2018). Re-examining the relationships among perceived quality, value, satisfaction, and destination loyalty: A higher-order structural model. *Journal of Vacation Marketing, 24*(2), 118–135. https://doi.org/10.1177/1356766717690572

Harrison-Walker, L. J. (2001). The measurement of word-of-mouth communication and an investigation of service quality and customer commitment as potential antecedents. *Journal of Service Research, 4*, 60–75.

Hikmah, H., Payangan, O. R., Munir, A. R., & Jusni, J. (2018). The effects of tourism products, service quality and destination uniqueness to the satisfaction and loyalty of tourist in South Sulawesi. *Scientific Research Journal, 6*(1), 91–100.

Kamrul, H., & Shamsul, K. A. (2019). Determining factors of tourists' loyalty to beach tourism destinations: A structural model. *Asia Pacific Journal of Marketing and Logistics, 32*(1), 169–187.

Kim, M., & Thapa, B. (2018). The influence of self-congruity, perceived value, and satisfaction on destination loyalty: A case study of the Korean DMZ. *Journal of Heritage Tourism, 13*(3), 224–236. https://doi.org/10.1080/1743873X.2017.1295973

Kohlmann, T. (2020). *Will coronavirus pandemic finally push emerging economies into crisis?* https://www.dw.com/en/coronavirus-economy/a-54168639

Kotler, P., & Keller, K. L. (2013). *Marketing management* (13th ed.). Pearson-Prentice Hall.

Lovelock, C. H. (2000). *Service marketing* (4th ed.). Prentice Hall.

Meleddu, M., Paci, R., & Pulina, M. (2015). Repeated behaviour and destination loyalty. *Tourism Management, 50*, 159–171.

Mensah, I., & Dei Mensah, R. (2013). *Management of hospitality and tourism services* (2nd ed.). Xlibris.

Mohamad, M., Ab Ghani, N. I., & Nasir, M. N. M. (2019). The impact of perceived value, destination service quality and tourist satisfaction on destination loyalty among international tourists visiting Malaysia. *Journal of Tourism, 4*(16), 10–26.

Narteh, B. (2018). Service quality and customer satisfaction in Ghanaian retail banks: The moderating role of price. *International Journal of Bank Marketing, 36*(1), 68–88. https://doi.org/10.1108/IJBM-08-2016-0118

Nguthi, S. T., Kambona, O., & Wadongo, B. (2021). Pull motivation factors influencing tourists' destination loyalty in the Lake Victoria Region tourism circuit, Kenya. *Journal of Hospitality and Tourism, 1*(1), 19–35.

Nguya, J., Kariuki, A., & Muthengi, S. (2021). Influence of tour guide service quality on customer satisfaction in Narok County, Kenya. *African Journal of Emerging Issues, 3*(1), 149–170.

Nilplub, C., Khang, D. B., & Krairit, D. (2016). Determinants of destination loyalty and the mediating role of tourist satisfaction. *Tourism Analysis, 21*(2–3), 221–236.

Oliver, R. L. (1997). *Satisfaction: A behavioral perspective on the consumer.* McGraw-Hill.

Oliver, R. L. (2006). Customer satisfaction research. In R. Grover & M. Vriens (Eds.), *The handbook of marketing research* (Chapter 27). Sage.

Parasuraman, A., Zeithaml, V. A., & Berry, L. L. (1988). SERVQUAL: A multiple-item scale for measuring consumer perception of service quality. *Journal of Retailing, 64*, 12–40.

Perić, G., Dramićanin, S., & Gašić, M. (2020). Impact of service quality on satisfaction and loyalty of tourists in rural tourism of Šumadija and Western Serbia. *Economics of Agriculture, 67*(4), 1071–1086. https://doi.org/10.5937/ekoPolj2004071P

Preko, A., Ameyibor, L. E. K., & Mohammed, I. (2020). Examining tour services, satisfaction and loyalty of international tourists in Cape Coast, Ghana. *Tourism Review International, 23*, 115–131. https://doi.org/10.3727/154427219X15741004672666

Raza, S. A., Umer, A., Qureshi, M. A., & Dahri, A. S. (2020). Internet banking service quality, e-customer satisfaction and loyalty: The modified e-SERVQUAL model. *The TQM Journal. Advance online publication.* https://doi.org/10.1108/TQM-02-2020-0019

Rubera, G., Ordanini, A., & Griffith, D. A. (2011). Incorporating cultural values for understanding the influence of perceived product creativity on intention to buy: An examination in Italy and the US. *Journal of International Business Studies, 42*(4), 459–476.

Sangpikul, A. (2018). The effects of travel experience dimensions on tourist satisfaction and destination loyalty: The case of an island destination. *International Journal of Culture, Tourism and Hospitality Research, 12*(1), 106–123.

Su, Y., Xu, J., Sotiriadis, M., & Shen, S. (2021). Authenticity, perceived value and loyalty in marine tourism destinations: The case of Zhoushan, Zhejiang Province, China. *Sustainability, 13,* 3716. https://doi.org/10.3390/su13073716

Tang, L., Scherer, R., & Morrison, A. M. (2011). Web site–based destination images: A comparison of Macau and Hong Kong. *Journal of China Tourism Research, 7*(1), 2–19. https://doi.org/10.1080/19388160.2011.551042

Veljković, S., & Đorđević, A. (2010). Brand values for consumers and companies. *Marketing, 41*(1), 3–16.

Wang, B., Yang, Z., Han, F., & Shi, H. (2017). Car tourism in Xinjiang: The mediation effect of perceived value and tourist satisfaction on the relationship between destination image and loyalty. *Sustainability, 9*(1), 22. https://doi.org/10.3390/su9010022

Wang, Y.-S., Wu, S.-C., Lin, H.-H., & Wang, Y.-Y. (2011). The relationship of service failure severity, service recovery justice and perceived switching costs with customer loyalty in the context of e-tailing. *International Journal of Information Management, 31*(4), 350–359.

Weru, J. N., & Njoroge, J. M. (2021). Investigating the influence of business events experience on international visitors' perceived destination image: The case of Kenya. *Journal of Convention & Event Tourism*, 1–23. https://doi.org/10.1080/15470148.2021.1895017

World Tourism Organization. (2011). *Tourism towards 2030/ Global overview.* UNWTO, Madrid. ISBN-13:978-92-844-1399-7

World Tourism Organization. (2017). *Tourism highlights (2017 ed.).* UNWTO, Madrid. https://doi.org/10.18111/9789284419029

World Tourism Organization. (2021). *International tourism highlights* (2020 ed.). UNWTO, Madrid. https://doi.org/10.18111/9789284422456

World Travel & Tourism Council. (2020). Travel & tourism economic impact 2019. https://wttcweb.on.uat.co/Research/Economic-Impact

Wu, H. C., & Li, T. (2015). An empirical study of the effects of service quality, visitor satisfaction, and emotions on behavioral intentions of visitors to the museums of Macau. *Journal of Quality Assurance in Hospitality & Tourism, 6*(1), 80–102.

Yeshi, S. (2019). The impact of quality of destination attributes on tourist satisfaction—The Case of Bhutan, Global. *Journal of Management and Business Research: B Economics and Commerce, 19*(5), 59–66.

7

Destination Distribution Systems and Travel Disintermediation in Emerging Economies

Christopher Mensah, Richmond Sakyi, and Mavis Adjoa Forson

Introduction

Until COVID-19 put a break on international and domestic travel, compelling the closure of hotels and attractions, as well as docking of cruise liners, tourism had witnessed considerable growth all over the world, contributing significantly to the gross domestic product of countries, creating employment for millions and generating much required foreign exchange for emerging economies in Africa, the Caribbean, Asia, and the Pacific regions. Increasing travel to emerging economies is significantly driving global growth in international tourism. The direct contribution of tourism to the gross domestic product of emerging economies such as Croatia, Mexico, Mauritius, Malaysia, and Morocco ranges between 15% and 7%

C. Mensah (✉) • R. Sakyi • M. A. Forson
Department of Hospitality and Tourism Management, Ho Technical University, Ho, Ghana
e-mail: cmensah@htu.edu.gh; rsakyi@htu.edu.gh; mforson@htu.edu.gh

I. Mensah et al. (eds.), *Marketing Tourist Destinations in Emerging Economies*, Palgrave Studies of Marketing in Emerging Economies,
https://doi.org/10.1007/978-3-030-83711-2_7

as of the year 2018 (United Nations World Tourism Organisation [UNWTO], 2020). Regarding the contribution to balance of payments, among emerging economies, Thailand and Macao boast of the largest travel surpluses, while Mexico, Croatia, Malaysia, India, the Dominican Republic, and Morocco recorded a surplus of US$6 billion to US$15 billion at the end of 2019 (UNWTO, 2020). It is worth noting that the growth of tourism in emerging economies in Southeast Asia and Africa has outperformed several regions including developed destinations in Europe and North America (Bunghez, 2020) on the account of providing tourists with exotic travel experiences in emerging tourist destinations.

Impactful tourism development hinges strongly on an effective and efficient tourism distribution system. A tourism distribution system is a web of intermediaries or middlemen who help suppliers and customers sell and distribute tourism services (Buhalis & Laws, 2001). Until recently, the influential intermediaries in the tourism distribution chain included traditional tour operators, travel agencies (TAs), charter brokers, and computer reservation systems. A combination of changes in the tourism business environment and the integration of information communication and technology in the operations of suppliers have resulted in the sidestepping of travel intermediaries in the distribution system, a phenomenon that has been described as disintermediation. The modernization of the tourism distribution system has attracted significant scholarly attention. However, the literature on tourism distribution systems in emerging economies remains fragmented. Consequently, this chapter explores the state of tourism distribution systems and online distribution intermediaries in emerging economies. In addition, online travel intermediation and disintermediation of traditional travel agencies in emerging economies are discussed.

Nature of Travel Intermediation in Emerging Economies

Travel intermediation is an important area of business operation within the travel and tourism industry because of its influence on the profitability and sustainability of tourism firms. Travel intermediation deals with the process by which tourism products and services are made available to consumers by third parties. In the tourism literature, travel intermediation is usually termed "tourism distribution system" or "tourism distribution channel". A tourism distribution channel is a network of intermediaries that make it easier for suppliers to sell and distribute tourism products and services to consumers (Kracht & Wang, 2010; Longhi, 2009). Travel intermediaries serve as middlemen in the travel industry, providing customers with travel details, hotel bookings, airline tickets, and car rental services (Lam & Zhang, 1998). The network of intermediaries constitutes independent organizations that work together to make a product or service accessible to customers or businesses.

Traditionally, three levels of the tourism distribution channel have been discussed in the literature (Fig. 7.1). The first level highlights a direct transaction between suppliers and consumers. At this level, tourism suppliers such as airlines and hotels make their products and services available to consumers through company computer reservation systems (CRS), sales offices located in regions, countries, and areas considered profitable markets. There are no third parties involved in the distribution

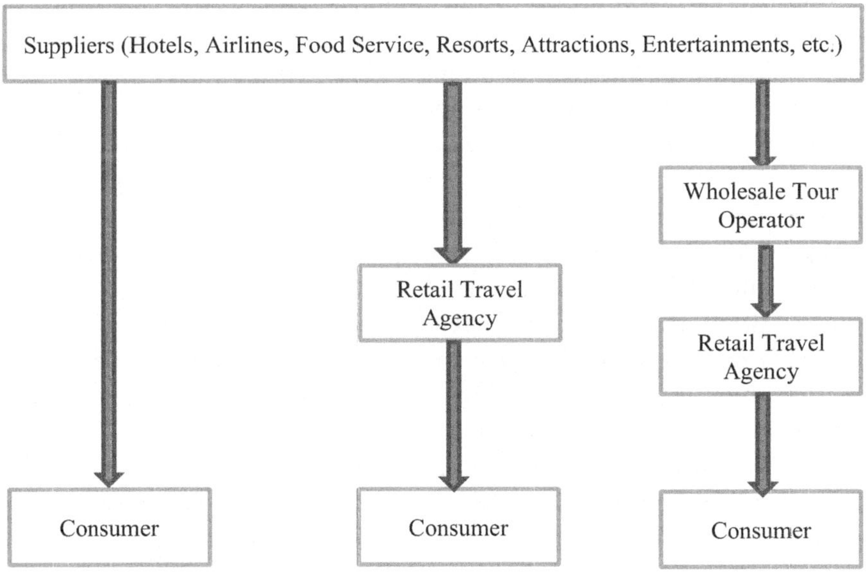

Fig. 7.1 Tourism distribution channel

of tourism services and products at this level. Prospective consumers make purchases or reservations via walk-ins, emails, or telephone. This first level of tourism distribution has also been described as "direct channel distribution" but in recent times, it has been referred to as "offline distribution" because of the growing influence of online distribution of tourism services and products. The suppliers of tourism services and products such as hotels and airlines sell directly to consumers via their websites (Barnett & Standing, 2001).

At the second and third levels of the tourism distribution channel, intermediaries are introduced. These are firms that play the role of retailers and wholesalers. Examples include travel agencies, tour operators, charter brokers, reservation systems, and travel distribution specialists (Buhalis, 2000). Within the travel and tourism sector, travel agencies and tour operators are the most popular travel intermediaries that dominate the distribution system in emerging economies. At the second level of the distribution channel, travel agencies playing the role of retail intermediaries between suppliers (airlines and hotels) and consumers and using global distribution systems (GDSs) such as Sabre, Galileo, Amadeus, and Worldspan earn commission from suppliers. Due to the emergence of online travel agencies (OTAs), the earlier models of travel agencies have been described in the literature using various terminologies such as "traditional travel agencies" or "brick and mortar" and "in-store" travel agencies. Travel agencies act as registered representatives and coordinators of the various components of the tourism product, selling their goods and services for commission. Again, travel agencies facilitate access to various tourist market segments and enhance providers' reputation by reducing quality uncertainty regarding the hotel product (Calveras & Orfila-Sintes, 2019). According to Wolfe et al. (2005), traditional travel agencies provide unique services that include extensive knowledge on travel matters, high integrity, and willingness to search for low fares as well as demonstrating courteous and friendly services.

The third level of the tourism distribution channel introduces a wholesaler, a role played by tour operators. A tour operator is a business that purchases individual travel components from suppliers and incorporates them into a package tour that is sold directly to the public or through middlemen with their own price tag. Tour operators are one of the most powerful and influential actors in the tourism industry (Alegre & Sard, 2017), determining the direction of tourism flows from the main source markets towards tourist destinations (Schwartz et al., 2008). Tour operators contact suppliers and travel agents, organize charter flights and pre-book accommodation, and market tour packages. During the trip, tour operators are responsible for local handling as well as dealing with local authorities and handling of special requests by guests (Buhalis & Ujma, 2006). Different categories of tour operators are discussed in the literature. These constitute specialist tour operators, incoming tour operators, and domestic tour operators. Specialist tour operators cater for special interest tourists during holidays, such as bird watching, cycling, photography, and golf (Fang, 2020). Incoming tour operators buy tourism services in their own countries and bundle them into products that are then promoted by partners in other countries. In addition, some incoming tour operators augment the services provided by outgoing tour operators by offering several optional services at the destination, and these include passenger transfers, tours, sporting programmes, and special events (Fang, 2020). Domestic tour operators compile and integrate the components of a tour into all-inclusive package for domestic travellers to buy. These tour operators usually provide travel services within the boundaries of their home country and sell travellers' package tours, such as domestic inclusive tours or independent tours. They usually target niche markets, such as the grey or youth market segments, and offer city vacations, coach holidays, or short breaks.

Current Trends in Travel Intermediation in Emerging Economies

Currently, the worldwide tourism distribution channel has undergone complex metamorphosis due to significant integration of information and communications technologies (ICTs) and the Internet in the travel and tourism industry, including emerging economies (Kracht & Wang, 2010). One major significant impact of ICT on travel intermediation is the emergence of OTAs or web-based travel agencies or virtual travel agencies. OTAs are travel aggregators that use the Internet to offer travel-related services such as flights, cruises, vacation packages, and hotel rooms to potential tourists (Talwar et al., 2020). OTAs market items from a variety of vendors, allowing customers to purchase the whole travel experience in one place (O'Connor & Frew, 2002). OTAs use the information gained from data mining to tailor direct mail promotions and loyalty programmes to the needs of customers (Toh et al., 2011). OTAs do not own properties but host websites with large collection of information, price comparisons, discounts, and review comments that attract tourists or business travellers (Chang et al., 2019). OTAs have attracted customers through the design of quality websites that assure clients of reliability, responsiveness, assurance, and empathy (Chang et al., 2019). Examples of OTAs include Booking.com, Expedia.com, Hotwire.com, Priceline, and Kayak.com. OTAs are gaining popularity among travellers. In 2012, they reported online sales of about US$100 billion (Carey et al., 2012) and almost tripled that figure, upward to US$278 billion, in 2013 (Knowles & Westcott, 2015).

The growth of OTAs started in the United States in the 1990s with the operations of Expedia.com, Travelocity.com, and Ortbiz.com (Leung et al., 2018) but has now spread to almost every part of the world, including emerging tourist destinations. Beyond the dominance of OTAs in Western countries, others have sprung up in the emerging economies. For example, there are MakeMyTrip in India, Trip.com in China, Wing of Travel in Hong Kong, HanaTour in Korea, and Jumia Travel in sub-Saharan Africa. In addition, Despegar.com operates in eight emerging economies as of 2015. These include Mexico, Argentina, Brazil, Colombia, Venezuela, Chile, Uruguay, and Peru (Majó et al., 2020). In China, the volume of online travel gross merchandise reached about US$90 billion in 2016. The three major OTA platforms in China (Ctrip, Tuniu, and Tongcheng) commanded a market share ranging from 43.6%, 22.7% to 11.1% respectively as at the end of December 2017 (Hou et al., 2019). The online travel market in India is estimated at US$9.1 billion in 2014 (Dutta et al., 2017). According to Irshad and Shakil (2019), the leading online travel markets in the Asia-Pacific region in terms of gross merchandise value are Thailand (US$4 billion), Indonesia (US$3 billion), and Vietnam (US$3 billion).

Several factors have driven the increasing patronage of OTAs in emerging economies. First, travellers derive many benefits from purchasing travel services online. One of the benefits includes pricing (Jedin & Ranjini, 2017). OTAs provide relatively cheaper rates for consumers, and this has driven the expanding share of OTAs in the travel market. Moreover, tourists find it more convenient using OTAs, as they can make flight and hotel room reservations in the comfort of their homes/locations at any time of the day (24/7 and 365 days) (Chakravarthi & Gopal, 2012; Shim et al., 2001). Increasing Internet penetration in emerging economies has contributed significantly to the growing popularity of online travel reservations through OTAs. Internet penetration in Korea, China, Thailand, Hong Kong, Brazil, Russian Federation, South Africa, and Mauritius, among many others, is on the increase, with some of the countries reporting Internet penetration of 96% (South Korea). The drive for online travel purchasing through OTAs has also been facilitated by improved digital skills and competency among the population of emerging economies. Enhanced digital skills have ensured that tourists feel self-efficacious to undertake online travel purchases. Other factors such as increasing accessibility and affordability of smartphones, laptop computers, and tablets, among others, with a corresponding deployment of software applications that allow easy purchasing of online travel services have all contributed significantly to the expansion of OTA operations in emerging economies. Other conditions influencing the use of OTAs include website serviceability (Lee et al., 2017; Lowry et al., 2008), flexible systems, information (Jeong et al., 2003), and traveller satisfaction (Law & Bai, 2008).

Role of Online Travel Intermediaries in the Marketing of Emerging Tourist Destinations

Marketing is cardinal to the attainment of competitive advantage, the success, and sustainability of destinations. Several stakeholders are involved in destination marketing and promotion, of which online travel intermediaries have become powerful stakeholders, following the digitalization of the tourism industry. Online travel intermediaries have global exposure and, therefore, benefit from the advantages of bulk selling. They provide a range of services, and these services are offered in more than one language, thereby allowing the customer to select a preferred language for a transaction. Besides, the full range of travel services offered by online travel intermediaries can easily be accessed via their websites. They also help in building brand reliability, as consumers can evaluate the benefits and form perceptions about the tourism product (Chivandi et al., 2020). Online intermediaries help tourism companies to gain exposure, as they have wider access to consumers. Through them, potential customers could search for products, prices, and directions, as well as encounters regarding services provided by suppliers (Rusu et al., 2016). With that, travel and tourism firms benefit from the "billboard effect" (Anderson, 2009). Some benefits derived by tourism service providers for using services from online travel intermediaries include advertisements, promotions, and direct transactions, as well as increased reservations obtained because of the billboard effect (Anderson, 2009; Majó et al., 2020). For example, OTAs in Malaysia assist SME hospitality operators with online advertising through content creation, dynamic banners, and tailored promos (Pacific Asia Travel Association [PATA], 2020).

Disintermediation in the Travel and Tourism Industry in Emerging Economies

Prior to the 1990s, the distribution of tourism services and products was solely undertaken by traditional travel agencies, tour operators, and other intermediaries. Airlines and hotels depended largely on travel agencies and tour operators to sell airline tickets, hotel room accommodation, and travel packages, among others, in exchange for commissions. However, the proliferation of information communication technologies (ICTs) and the Internet, coupled with the strategic decisions of airlines, hotels, and other suppliers to reduce distribution cost, led to a considerable cut in commissions paid to travel intermediaries. In addition, the strategic decision to cut commissions paid to travel agencies was necessitated by health catastrophes and consequences of safety concerns due to military actions. For example, hotel and airline operators in Hong Kong were reported to have cut commissions to travel agencies in the early 2000s due to health and military actions (Tse, 2003). In the 1990s, many airlines reduced the commissions paid to travel agents and added fees to force consumers to directly purchase tickets from the suppliers (Buhalis & Moldavska, 2021).

To reduce distribution costs, airlines and hotels in emerging economies, similar to their counterparts in the Americas and Europe, have taken advantage of the increasing global access to the Internet to design corporate websites that provide opportunities for prospective clients to directly purchase their services without patronizing the services of intermediaries. Growing Internet access, availability of ICT infrastructure (smartphone, tablets, laptop computers), software applications for online reservations, and growing technology-savvy travellers have ensured that tourism service providers in emerging countries, such as Hong Kong, Thailand, Brazil, Singapore, Korea, Vietnam, and Malaysia, are able to sell directly to prospective tourists through their websites and without recourse to intermediaries. For example, low-cost carriers, such as Air Asia, Lion Air, and Cebu Pacific in Southeast Asia, provide an opportunity for travellers to make ticket reservations and payment through their websites. Shangri-La hotels in Hong Kong and China make provision for customer reservation through the hotel website. This situation has led to disintermediation in the chain of the tourism distribution system.

There are several definitions of disintermediation in the literature. Disintermediation is the removal of intermediaries from a process, supply chain, or market. It is the elimination of intermediaries within the distribution channel, propelled by electronic means that enable consumers to access and transact directly with suppliers and destinations (Bennett & Buhalis, 2003). Disintermediation in the distribution of tourism services is marked by a reduction in the market share of traditional travel agencies (Pötzl, 2000; Tsai et al., 2005). In the view of Kaewkitipong (2010), disintermediation is the idea of eliminating middlemen/mediators from the supplier-intermediary-consumer network. According to Law et al. (2015), disintermediation refers to the disengagement of traditional travel agencies (i.e. offline travel agencies and tour operators who help distribute tourism products and/or services from suppliers to consumers) from the supplier-intermediary-consumer network. Disintermediation within the travel and tourism distribution system could also be attributed to strategies of suppliers to distribute their products and services using evolving online technologies, such as social media and mobile technologies (Thakran & Verma, 2013). TAs in Brunei have faced disintermediation because most travel suppliers (hotels and airlines) offer web-based reservation systems that facilitate direct purchases. Furthermore, the majority of independent buyers are connected to the Internet (Almunawar et al., 2013) and are, therefore, able to make reservations at the comfort of their homes without requiring the assistance of intermediaries.

Disintermediation in the distribution of tourism services is marked by a reduction in the market share of traditional travel agencies (Pötzl, 2000; Tsai et al., 2005). There have been reports of decline in the number of travel intermediaries globally due to disintermediation. For example, in the mid-1990s, there were over 34,000 travel agents in the United States; however, the number reduced to 13,000 by the year 2013 (Cable News Network [CNN], 2013). Similarly, the role and share of travel agencies in the distribution of travel services and products declined in emerging economies due to competition from online travel agencies and supplier websites. For example, in Malaysia, the number of registered traditional travel agencies reduced from 4691 in early 2000 to 3594 as of 2019 (Fuza et al., 2015; Malaysian Association of Tour and Travel Agents [MATTA], 2019).

Challenges and the Future of Travel Intermediation in Emerging Economies

Given the important role of travel intermediation in the long-term sustainability of emerging tourist destinations, it is important to shed light on the challenges that confront the tourism distribution channel. According to Buhalis (2003), the participation of OTAs in the travel distribution channel has created several challenges relating to marketing and sales (O'Connor & Frew, 2004). Interfirm conflicts within the tourism value chain have been documented in the literature (Buhalis, 2000; Law et al., 2015; Tan & Dwyer, 2014). The tourism industry is a coordination-intensive industry, and this demands collaboration and cooperation among the various stakeholders for quality service delivery. Conflicts among the various collaborators have the potential to affect profitability. It has been reported that tour operators have the tendency to jeopardize the image, profit margins, and long-term profitability of airlines and hotels as a means of increasing their profits (Bastakis et al., 2004). Another source of interfirm conflict results from the provision of lower quality services by hotels to clients of tour operators than what was initially promised (Candela & Figini, 2012). These behaviours constitute grave sources of conflicts in the tourism distribution channel. There are also reports of conflicts between hotels and OTAs due to concerns about rate parity and commissions (Lee et al., 2013).

Future of Travel Intermediation in Emerging Economies

The future survival of the tourism industry in a post-COVID-19 era is partially dependent on the exploitation, usage, and management of ICT. The emergence of different types of online travel intermediaries and the design of attractive and interactive supplier websites with 11th-hour deals present unique opportunities for tourists to access tourism commodities (Law et al., 2014, 2015). Not much is said about the future extinction of traditional travel intermediaries, although a lot has been said concerning the recent increase in ICT and its effect on travel intermediation (disintermediation) (Law et al., 2014, 2015). However, some researchers have predicted the future extinction of traditional travel intermediaries, while others still believe they will continue to be important and will still be in operation in the future (Hills & Cairncross, 2011; Law et al., 2014, 2015).

In view of the extent to which ICT is permeating the tourism industry, it is assumed that the need for consumers to walk directly to the travel agency would disappear in the future as the Internet readily provides tourism information and destinations for consumers (Law et al., 2014, 2015; Morosan & Jeong, 2008; Piccoli & Lloyd, 2010). Also, the difficulty in hunting for tourism information is now reduced by the existence of the Internet and ICT; therefore, consumers can obtain tourism information regardless of location, day, and time (Kracht & Wang, 2010; Law et al., 2014, 2015; Yacouel & Fleischer, 2012).

Incidentally, the Internet provides a wider variety of options to consumers and will allow consumers to make relevant decisions as compared to traditional travel intermediation (Law et al., 2014, 2015; Wang & Wang, 2010). Through the Internet, all kinds of reservations can be made by consumers, and special promotional deals and discounts will be available to consumers who purchase products from Internet pages of providers (Law et al., 2014, 2015). Most reservations, as stated by Grønflaten (2009), are made through the Internet using online intermediaries. Therefore, the future will see enhancement of Internet facilities and usage by older generation of consumers (Prayag & Del Chiappa, 2014). Tourism providers will operate in different countries, as more tourists will travel from emerging economies of the world. Also, new electronic systems will be introduced in the tourism industry where travel intermediaries will operate both physically and online (Camilleri, 2018).

However, travel intermediaries will still be important in destination marketing since the provision of personalized services and face-to-face contacts between tourists and intermediaries offer the tourist a sense of security, privacy, and trust. It also gives better opportunities to tourists and is relevant to some aspects of the market, including packaged tour consumers, business tourists, tourists with low or no educational qualification and digital skills, and consumers interested in customized packages (Del Chiappa, 2013; Law et al., 2014, 2015). In the future, the specialized skills and expertise of traditional travel intermediaries cannot be substituted for the Internet, but travel intermediaries will be required to provide customized and attractive services that meet the specific needs of customers (Camilleri, 2018). Although the Internet supports reservations among other services and while its usage might increase in the future, some tourists will still patronize travel intermediaries for expert information and assistance (Law et al., 2014, 2015).

Conclusion

In this chapter, we explored the state of travel distribution systems, online intermediaries and disintermediation in emerging economies. Clearly, travel distribution systems in emerging destinations have undergone considerable metamorphosis due to the extensive integration of ICTs in the operations of travel service providers. This has also been complemented by increasing Internet penetration and improving digital literacy among the population of emerging destinations. This situation has enhanced the increasing prominence of OTAs in travel service distribution as well as the development of efficient service provider websites in emerging destinations. Indigenous OTAs have emerged in China, India, Hong Kong, Korea, and Thailand, among others. These indigenous OTAs are driving the increasing online distribution of travel services. Consequently, disintermediation has become common in the travel distribution landscape in emerging destinations as tourism service providers are selling directly to clients through their interactive and easy-to-use websites. The share of traditional travel agencies in the distribution of travel services has declined considerably in emerging tourist destinations. With the devastating consequences of COVID-19 on the hospitality and tourism industry, online travel intermediaries are key to the restart and recovery of the sector.

References

Alegre, J., & Sard, M. (2017). Price strategies by German and British tour operators in Mallorca. *Journal of Hospitality and Tourism Management, 33*, 93–102.

Almunawar, M. N., Anshari, M., & Susanto, H. (2013). Crafting strategies for sustainability: How travel agents should react in facing a disintermediation. *Operational Research, 13*(3), 317–342.

Anderson, C. (2009). The billboard effect: Online travel agent impact on non-OTA reservation volume. *Cornel Hospitality Report, 9*(16), 4–10.

Barnett, M., & Standing, C. (2001). Repositioning travel agencies on the Internet. *Journal of Vacation Marketing, 7*(2), 143–152.

Bastakis, C., Buhalis, D., & Butler, R. (2004). The perception of small and medium sized tourism accommodation providers on the impacts of the tour operators' power in Eastern Mediterranean. *Tourism Management, 25*(2), 151–170.

Bennett, M., & Buhalis, D. (2003). *Tourism and travel distribution: The travel agent's perspective.* Centre for e-tourism Research, School of Management, University of Surrey.

Buhalis, D. (2000). Relationships in the distribution channel of tourism: Conflicts between hoteliers and tour operators in the Mediterranean region. *International Journal of Hospitality & Tourism Administration, 1*(1), 113–139.

Buhalis, D. (2003). *eTourism: Information technology for strategic tourism management.* Pearson Education.

Buhalis, D., & Laws, E. (2001). *Tourism distribution channels: Practices, issues and tourism distribution channels: Practices, issues and transformations.* Continuum.

Buhalis, D., & Moldavska, I. (2021). In-room voice-based AI digital assistants transforming on-site hotel services and guests' experiences. In *Information and communication technologies in tourism 2021* (pp. 30–44). Springer.

Buhalis, D., & Ujma, D. (2006). Intermediaries: Travel agencies and tour operators. In D. Buhalis & C. Costa (Eds.), *Tourism Business Frontiers* (pp. 171–180). Elsevier Buttterworth-Heinemann.

Bunghez, C. L. (2020). Marketing strategies of travel agencies: A quantitative approach. *Sustainability, 12*(24), 10660.

Calveras, A., & Orfila-Sintes, F. (2019). Intermediation in hospitality and transaction cost theory: Evidence from the Balearic Islands, 2001–2010. *Journal of Destination Marketing & Management, 11*, 281–291.

Camilleri, M. A. (2018). Tourism distribution channels. In *Travel marketing, tourism economics and the airline product* (pp. 105–115). Springer.

Candela, G., & Figini, P. (2012). The economics of tourism destinations. In *The economics of tourism destinations* (pp. 73–130). Springer.

Carey, R., Kang, D., & Zea, M. (2012). The trouble with travel distribution. *McKinsey Quarterly*. http://mckinseyquarterly.com/transportation/strategy

Chakravarthi, J. S. K., & Gopal, V. (2012). Comparison of traditional and online travel services: A concept note. *IUP Journal of Business Strategy, 9*(1), 45–58.

Chang, Y. W., Hsu, P. Y., & Lan, Y. C. (2019). Cooperation and competition between online travel agencies and hotels. *Tourism Management, 71*, 187–196.

Chivandi, A., Samuel, O., & Muchie, M. (2020). The effect of online travel agencies (OTA) on brand relationships in low and high-end hotels in South Africa. *Geo Journal of Tourism and Geosites, 31*(3), 951–957.

CNN. (2013). *The travel agent is dying, but it's not yet dead.* https://edition.cnn.com/travel/article/travel-agent-survival/index.html

Del Chiappa, G. D. (2013). Internet versus travel agencies: The perception of different groups of Italian online buyers. *Journal of Vacation Marketing, 19*(1), 55–66.

Dutta, S., Kumar, C. R., & Chauhan, K. (2017). Factors affecting customer satisfaction of online travel agencies in India. *Tourism and Hospitality Management, 23*(2), 267–277.

Fang, W. T. (2020). *Tourism in Emerging Economies*. Springer.

Fuza, Z. I. M., Ismail, H. N., & Isa, S. M. (2015). Importance of information communication technologies (ICT) among tour operators in Malaysia. *Advanced Science Letters, 21*, 1650–1654.

Grønflaten, Ø. (2009). Predicting travelers' choice of information sources and information channels. *Journal of Travel Research, 48*(2), 230–244.

Hills, J. R., & Cairncross, G. (2011). How small regional accommodation providers view and respond to online intermediaries. *Journal of Vacation Marketing, 17*(4), 249–262.

Hou, Z., Cui, F., Meng, Y., Lian, T., & Yu, C. (2019). Opinion mining from online travel reviews: A comparative analysis of Chinese major OTAs using semantic association analysis. *Tourism Management, 74*, 276–289.

Irshad, M., & Shakil, A. (2019). Investigating the determinants of consumers' attitude towards social media marketing: Moderating role of gender. *Online Journal of Communication and Media Technologies, 9*(4), 1–15.

Jedin, M. H., & Ranjini, K. (2017). Exploring the key factors of hotel online booking through online travel agency. In *4th International Conference on E-Commerce (ICoEC) 2017 Held in Malaysisa.*

Jeong, M., Oh, H., & Gregoire, M. (2003). Conceptualizing web site quality and its consequences in the lodging industry. *International Journal of Hospitality Management, 22*, 161–175.

Kaewkitipong, L. (2010, August). Disintermediation in the tourism industry: Theory vs. practice. In *SIGeBIZ track of the Americas Conference on Information Systems* (pp. 160–171). Springer.

Knowles, H., & Westcott, M. (2015). *Travel services.* https://ecampusontario.pressbooks.pub/introtourismbc/chapter/chapter-7-travel-services/

Kracht, J., & Wang, Y. (2010). Examining the tourism distribution channel: Evolution and transformation. *International Journal of Contemporary Hospitality Management, 22*(5), 736–757.

Lam, T., & Zhang, H. Q. (1998). Service quality of travel agents: The case of travel agents in Hong Kong. *Tourism Management, 20*(3), 341–349.

Law, R., & Bai, B. (2008). How do the preferences of online buyers and browsers differ on the design and content of travel websites? *International Journal of Contemporary Hospitality Management, 20*(4), 388–400.

Law, R., Buhalis, D., & Cobanoglu, C. (2014). Progress on information and communication technologies in hospitality and tourism. *International Journal of Contemporary Hospitality Management, 26*(5), 727–750.

Law, R., Leung, R., Lo, A., Leung, D., & Fong, L. H. N. (2015). Distribution channel in hospitality and tourism. *International Journal of Contemporary Hospitality Management, 27*(3), 431–452.

Lee, H. A., Denizci, G. B., & Law, R. (2013). An examination of the relationship between online travel agents and hotels: A case study of Choice Hotels International and Expedia.com. *Cornell Hospitality Quarterly, 54*(1), 95–107.

Lee, M. K., Yoon, H. Y., & Park, H. W. (2017). From online via offline to online: How online visibility of tourism information shapes and is shaped by offline visits. *Journal of Travel & Tourism Marketing, 34*(9), 1143–1154.

Leung, R., Au, N., Liu, J., & Law, R. (2018). Do customers share the same perspective? A study on online OTAs ratings versus user ratings of Hong Kong hotels. *Journal of Vacation Marketing, 24*(2), 103–117.

Longhi, C. (2009). Internet and organisation of the industry in tourism: A focus on the distribution of travel and tourism services. *International Journal of Leisure and Tourism Marketing, 1*(2), 131–151.

Lowry, P. B., Vance, A., Moody, G., Beckman, B., & Read, A. (2008). Explaining and predicting the impact of branding alliances and web site quality on initial consumer trust of e-commerce web sites. *Journal of Management Information Systems, 24*(4), 199–224.

Majó, F. J. M., Vall-Llosera, L., & Moya, D. H. (2020). Analysis of OTA impact on hotel reservations. Case study: Bogotá. *Turismo y Sociedad*, 145–159. https://doi.org/10.18601/01207555.n27.08

Malaysian Association of Tour and Travel Agents. (2019). *Members directory.* https://www.matta.org.my/members.

Morosan, C., & Jeong, M. (2008). Users' perceptions of two types of hotel reservation Web sites. *International Journal of Hospitality Management, 27*(2), 284–292.

O'Connor, P., & Frew, A. J. (2002). The future of hotel electronic distribution: Expert and industry perspectives. *Cornell Hotel and Restaurant Administration Quarterly, 43*(3), 33–45.

O'Connor, P., & Frew, A. J. (2004). An evaluation methodology for hotel electronic channels of distribution. *International Journal of Hospitality Management, 23*(2), 179–199.
Pacific Asia Travel Association. (2020). *The role of online travel agencies in supporting Asia Pacific SMEs in recovery*. Pacific Asia Travel Association.
Piccoli, G., & Lloyd, R. (2010). Strategic impacts of IT-enabled consumer power: Insight from Internet distribution in the US lodging industry. *Information & Management, 47*(7–8), 333–340.
Pötzl, J. (2000). Issues in direct channel distribution: A comparison of selling via the internet in the airline business and the fast-moving consumer goods industry. *Electronic Markets, 10*(3), 153–157.
Prayag, G., & Del Chiappa, G. (2014). Hotel disintermediation in France: Perceptions of students from Generation Y. *Anatolia, 25*(3), 417–430.
Rusu, V., Rusu, C., Guzmán, D., Espinoza, D., Rojas, D., Roncagliolo, S., & Quiñones, D. (2016). Assessing the customer experience based on quantitative data: Virtual travel agencies. In *International Conference of Design, User Experience, and Usability* (pp. 499–508). Springer.
Schwartz, K., Tapper, R., & Font, X. (2008). A sustainable supply chain management framework for tour operators. *Journal of Sustainable Tourism, 16*(3), 298–314.
Shim, S., Eastlick, M. A., Lotz, S. L., & Warrington, P. (2001). An online prepurchase intentions model: The role of intention to search: Best overall paper award—The Sixth Triennial AMS/ACRA Retailing Conference, 2000☆. *Journal of Retailing, 77*(3), 397–416.
Talwar, S., Dhir, A., Kaur, P., & Mäntymäki, M. (2020). Why do people purchase from online travel agencies (OTAs)? A consumption values perspective. *International Journal of Hospitality Management, 88*, 102534.
Tan, J., & Dwyer, L. (2014). Managing distribution channel conflict in the hotel industry. *International Journal of Marketing & Business Communication, 3*(1), 9–19.
Thakran, K., & Verma, R. (2013). The emergence of hybrid online distribution channels in travel, tourism and hospitality. *Cornell Hospitality Quarterly, 54*(3), 240–247.
Toh, R. S., Raven, P., & DeKay, F. (2011). Selling rooms: Hotels vs. third-party websites. *Cornell Hospitality Quarterly, 52*(2), 181–189.
Tsai, H. T., Huang, L., & Lin, C. G. (2005). Emerging e-commerce development model for Taiwanese travel agencies. *Tourism Management, 26*(5), 787–796.

Tse, A. C. B. (2003). Disintermediation of travel agents in the hotel industry. *International Journal of Hospitality Management, 22*(4), 453–460.
UNWTO. (2020). *International tourism highlights*. United Nations World Tourism Organization.
Wang, H. Y., & Wang, S. H. (2010). Predicting mobile hotel reservation adoption: Insight from a perceived value standpoint. *International Journal of Hospitality Management, 29*(4), 598–608.
Wolfe, K., Hsu, C. H., & Kang, S. K. (2005). Buyer characteristics among users of various travel intermediaries. *Journal of Travel & Tourism Marketing, 17*(2–3), 51–62.
Yacouel, N., & Fleischer, A. (2012). The role of cybermediaries in reputation building and price premiums in the online hotel market. *Journal of Travel Research, 51*(2), 219–226.

8

The Progress of ICT Applications in Destination Marketing in Emerging Economies

Kandappan Balasubramanian
and S. Mostafa Rasoolimanesh

Introduction

The growth of many emerging economies has been predominantly driven by the tourism industry. Many countries are competing to attract potential tourists with the advancement of information and communications technology (ICT), as it creates many new opportunities to open the market (Aslan et al., 2021; Almatourism, 2019). There are some fundamental changes in tourism products and services due to the acceleration and interaction between ICT and tourism that have an influence on tourist behaviour (Chen, 2020). The ICT revolution has had profound changes in the lives of people, organisations, and the positioning of the tourism destinations with the promotion of tourist

K. Balasubramanian (✉) • S. M. Rasoolimanesh
Taylor's University, Selangor, Malaysia
e-mail: kandappan.balasubramanian@taylors.edu.my;
mostafa.rasoolimanesh@taylors.edu.my

I. Mensah et al. (eds.), *Marketing Tourist Destinations in Emerging Economies*, Palgrave Studies of Marketing in Emerging Economies,
https://doi.org/10.1007/978-3-030-83711-2_8

attractions and products (Wagaw & Mulugeta, 2018). Information is the lifeblood of tourism; hence, timely and accurate information that is relevant to customers is key to tourism production and satisfaction of demand (Buhalis, 2003). One of the significant forces driving change in the tourism industry is information technology. Since the year 2000, people have been witnessing the real transformational effect of ICTs and the Internet. ICTs, thus, play a critical role in the competitiveness of tourism organisations and destinations, as well as in the entire industry as a whole (Buhalis, 2003).

The use of ICT applications has become the norm during the COVID-19 era with all the social distance restrictions. ICT innovations have altered not only the ways people search for information about destinations and purchase travel products but also how they experience, communicate, and perceive destinations (Buhalis, 2003). The proliferation of ICT, with particular emphasis on the Internet, presents new opportunities for the marketing of emerging destinations. The e-tourism concept involves all business operations, from e-marketing, e-commerce, and e-finance to e-procurement, e-accounting, e-R&D, and e-HRM (Buhalis, 2003). E-marketing is the use of information technology in the processes of creating, communicating, and delivering value to customers, and for managing customer relationships in ways that benefit the organisation and its stakeholders.

The tourism industry is a dynamic service sector that faces tough competition globally, with constant changes in customers' demands and expectations. That is why innovation is important for the survival of destinations and tourism businesses. As in other sectors such as manufacturing, the survival of tourism companies depends on their innovation. Otherwise, their services or products may become absolute in the market. With the support of stakeholders' collaboration, the entrepreneurs can play a significant role through exploiting global opportunities to achieve the branding destination image (Burns et al., 2014; Del Chiappa & Presenza, 2013; Shams, 2016). Investing in ICT to improve customer services is always considered an innovation process in the tourism industry (Labanauskaitė et al., 2020). The scope of this chapter relates to the effects of ICTs on destination marketing in emerging economies. In the first part of the chapter, the nature and scope of e-tourism and v-tourism are presented. The chapter also elaborates on different components of e-tourism and v-tourism and their application to destination marketing in emerging economies, including destination management systems (DMSs), e-marketing of tourist destinations, as well as augmented reality (AR) and virtual reality (VR) for the promotion of emerging destinations. The chapter also discusses the challenges and opportunities of ICT utilisation for destination marketing in emerging economies.

Social Media Marketing (SMM)

Social media usage in the tourism industry has increased rapidly beyond the horizons of information technology (Sarkar & George, 2018). Social media have enabled many travellers to schedule their journeys in advance through accessing the enormous available travel information, such as picture- and video-sharing pages, wikis, blogs, and online review sites (Hua et al., 2017). Social networking is one of the Internet environment's fastest-growing communication technologies. Social media marketing refers to online media such as Facebook, YouTube, Instagram, Twitter, Ustream, and TripAdvisor in which people create their own personal profiles to share their interests, experiences, and information for social interactions (Cutshall et al., 2021). The most important role of social media (SM) is to generate and upload a wide range of content in the form of text, images, videos, and so on, through a web-based application (Kang & Schuett, 2013). Social networking itself is an all-round concept for platforms that can have social activities that are fundamentally new. For example, Twitter is a networking website that allows users to exchange brief or updated messages with others. In comparison, Facebook is a comprehensive social networking platform that enables alerts, photographs, event attendances, and a number of other activities to be shared (Garg & Pahuja, 2020). Social media technology enables travellers to share different sets of content easily and comfortably to impact future travellers' decision-making. Social media identity and internationalisation were considered key factors in creating perceived pleasure among travel experiences while accessing the destination information (Kavoura & Stavrianea, 2014).

E-Tourism: An Overview

The development of technology and applications during the digitalisation era of IR4.0 is changing lives every day by opening up new opportunities, which also transform the tourism business globally (OECD Tourism Trends and Policies, 2020). Due to digitalisation and emerging new technologies, billions of people worldwide consider travel and tourism as the main leisure activity which opened the door to e-tourism (Wanderer, 2019). One of the well-known definitions for e-tourism is "e-tourism reflects the digitalisation of all processes and value chains in the tourism, travel, hospitality and catering industries" (Buhalis, 2003, p. 76). E-tourism encompasses the analysis, design, and implementation of information technology/e-commerce solutions to the travel and tourism industry, as well as the analysis of the effects of the various technical/economic processes and market structures. Over the years, the e-tourism concept has been defined by various authors differently, as shown in Table 8.1:

Table 8.1 shows that some authors characterise "e-tourism" as "not only electronic distribution of tourist services, but also electronic

Table 8.1 Descriptions of e-tourism

Description	Authors (year)
Not only electronic distribution of tourist services but also electronic excursions, which are also called virtual	Moshnjaga (2013)
The widespread use of ICT opportunities for a full cycle of business activity	Kalmakova (2015)
Study of tourism manifestations by ICT broadly	Jingjin Li et al. (2018)
ICT in tourism management	Navío Marco et al. (2018)
Just one of the outcomes of the tourism industry's incorporation of technology	Sanaz Shaee et al. (2019)
Due to computing and pervasive connectivity, technology today is no longer just a tool for e-tourism but is used in all aspects of life and travel	
A part of e-commerce that integrates rapidly developing areas such as telecommunications and information technology into the hospitality and management industry	Shevchenko et al. (2019)

Source: Olga Kononova et al. (2020)

excursions, which are also called virtual" (Moshnjaga, 2013), noting that the appearance of this term is associated "with the transformation of the term e-business, which is the use of a wide range of ICT capabilities for organising a full cycle of business activities (e-commerce, e-marketing, e-finance, e-production, e-strategy, e- management)" (Kalmakova, 2015). The phenomenon is "part of e-commerce and integrates rapidly developing areas such as telecommunications and information technology into the hospitality and tourism industry" (Shevchenko et al., 2019).

The development of e-tourism is basically determined by the interactions between the tourism organisation, consumer, and the government with the changing consumer behaviour due to technology innovation (Nedelea & Bălan, 2010). E-tourism is essentially the digitalisation of the whole touristic industry and infrastructure. Some of the advantages of e-tourism are reduction of seasonality, more successful communication with tourists and an increase in reservations and sales. The use of the Internet has forever changed the structure and the principles of the tourism industry. Tourists are now capable of easily choosing their destinations, comparing prices, and managing their financial exchanges. Information and communications technologies (ICTs), and the Internet, if wisely used, could prove to be highly innovative strategic tools in the hands of tourism operators, which could help improve their competitive positions (Pitoska, 2013).

The e-Tourism System

The suggested e-tourism system in Fig. 8.1 combines leading concepts into a complex and dynamic e-tourism system based on four significant groups of elements (e-inputs, web-marketing subsystem, destination subsystem web services, and e-outputs) to have a paradigm shift in the travel and tourism industry. This shift will bring a thorough change to the structure and way the contemporary travel and tourism industry functions and keeps providing both new utilities and opportunities, as well as new threats and challenges at the same time. E-tourism growth has key effects on the competitive abilities of travel organisations and destinations. The successful functioning of the e-tourism system requires a

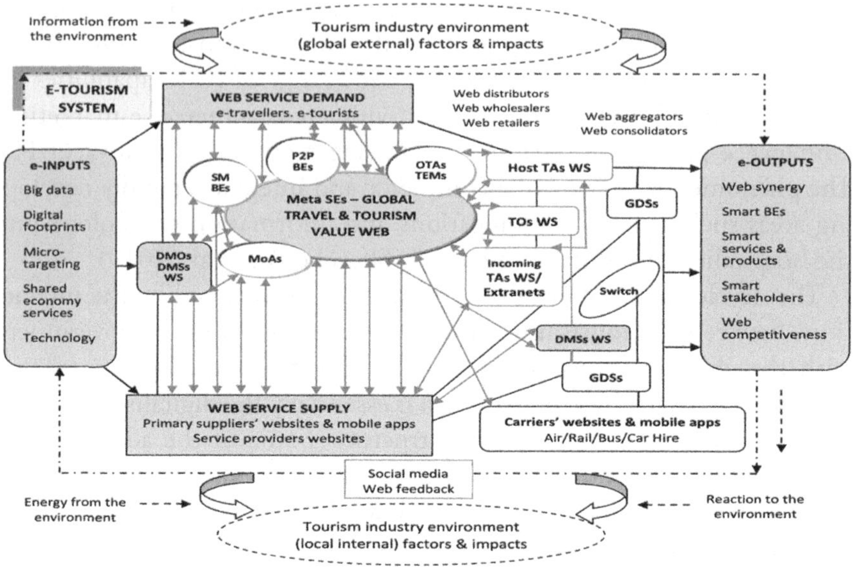

Fig. 8.1 E-tourism system's conceptual framework (Kazandzhieva & Santana, 2019). (Notes: DMOs: Destination management organisations; DMSs: destination management systems [TISCover, VisitSingapore, VisitFinland, MySwitzerland, VisitBritain]; WS: web service; SM BEs: social media booking engines [TripAdvisor, HolidayCheck, Facebook, Twitter]; MoAs: mobile applications; P2P BEs: per-to-per booking engines [Airbnb, Uber]; Meta SEs: metasearch engines [Google, Kayak, Skyscanner]; OTAs/TEMs: online travel agencies/tourism e-mediaries [Expedia, Orbitz, Travelocity]; TAs WS: travel agents web service; TOs WS: tour operators web service; GDSs: global distribution systems [Amadeus, Galileo]. Source: Velina Kazandzhieva/Hristina Santana, E-tourism: Definition, development and conceptual framework)

proactive, innovative management approach, leading to constant organisational and technological integration and development.

E-inputs—incoming means and resources. During the past decade, there has been enormous information emanating from the Big Data within the exchange flow of e-tourism systems. For tourism management, in the digital era, Big Data represents efficient real-time analysis for optimal use of information to improve e-tourism and supply intelligent travel experience to customers. All e-travellers leave digital footprints (active or passive) on the web. Accessing uploaded photos, online reviews, or clicking on links presents an entirely new and effective tool for collecting actual data and exploring e-demand and visitor flow management (Olmedo et al., 2018).

Web-marketing subsystem. It comprises two parts of the digital tourism market—web service supply and demand. The system is characterised by significant transformations in the online distribution channels and their diversification; the emergence of new players and traditional travel intermediaries' dropping off; and direct and bidirectional, accelerated, and facilitated communication with end consumers and primary suppliers, thanks to the peer-to-peer booking platforms such as Airbnb and Uber (Kazandzhieva & Santana, 2019). A significantly important transformation in the online distribution system in travel and tourism is the process of interconnecting social media with OTAs/TEMs and the primary suppliers' websites (Lee et al., 2013).

Destination subsystem web service is based on DMOs and their networks of interaction with tourism boards, governmental and non-governmental bodies, professional organisations, primary suppliers, and service providers at different levels (local, regional, national). The subsystem has a major impact on SMEs' active participation in the e-tourism business. Connecting DMSs' web service with mobile apps of meta SEs and SM booking platforms personalises travel and tourism and improves the experience by offering individual local services and products. With better integration of the technological structure of the DMOs in the e-tourism system, many favourable facilities to utilise the potential of the multiple online channels are being created (Kazandzhieva & Santana, 2019).

E-outputs—Outgoing resources. Cumulative results and effects are obtained through a combination of the subsystems and various trends and influences in e-tourism development. These results come into the form of synergy effects, basically in relation to the expansion in the range of the e-tourism system; balanced and coordinated development with added value and benefits; digital innovation to all stakeholders; strategic effectiveness and profitability; and sustainable competition stimulation. Smart services and products supplied with suitable intelligent booking engines increase the e-tourism system's effectiveness in both the destinations and the local tourism business market. Implementing modern technology allows the transformation of substantial data quantity into web value for travellers and tourists (Kazandzhieva & Santana, 2019).

The digitalisation of the traditional travel and tourism system and its transformation into the e-tourism system are not just a virtual future, but an actual reality. Many more destinations and researchers are now focusing on e-tourism. The proposed model of an e-tourism system, as shown in Fig. 8.1, is a paradigm shift that brings a thorough change to the structure and way the contemporary travel and tourism industry functions and keeps providing both new utilities and opportunities as well as new threats and challenges at the same time. E-tourism growth has a key effect on the competitive abilities of travel organisations and destinations. The successful functioning of the e-tourism system requires a proactive, innovative management approach leading to constant organisational and technological integration and development. The e-tourism approach is an opportunity to fully exploit its potentials and scope, a strategic purpose to maintain companies' and destinations' positions in a digital economy. Otherwise, they will suffer from adaptation difficulties and restrictions to keep their positions in the e-tourism system and the liability of dropping off. The dynamics in the development of e-tourism requires quick and smart arrangements from all stakeholders. In the uncertain future of the e-tourism system environment, companies have to clearly define their (market) niche or tourism specialisation, know their needs and desires (personalisation), highlight the exact competitive advantages, and use them effectively (differentiation). At the same time, tourism service providers need to be flexible and adaptable to the changes within the external and business environment (Kononova et al., 2020).

The Transformation from E-Tourism to V-Tourism Using Virtual and Augmented Reality in the Tourism Sector

The latest advancement in VR and AR has brought about a new industrial revolution for the tourism industry, which creates a new experience through travel- and tourism-related products and services. The development of new technologies is transforming the physical and virtual spaces by opening new markets with the integration of VR, AR, and mixed reality. Those changes have brought about a new paradigm shift in the marketing strategies of the tourism industry to enhance customer experiences and become innovative players (Egger & Neuburger, 2020). VR plays an important role in positioning sustainable destinations and updating the tourism industry with information in today's fast-paced industry. The evolution of the smartphone and enhanced mobile has transformed tourists' behaviour towards accessing and sharing information before, during, and after their travels (Jung et al., 2015). The traditional destination experience with the tour guides was replaced with the context-aware tourism experience due to the increased popularity of smartphone and the development of AR applications. The integration features such as an in-built camera, a global positioning system (GPS), and Internet connection available in all smartphones replace the role of tour guide for any destination direction and information (Chou & Chanlin, 2012; Yovcheva et al., 2013).

AR is defined as "a technology that allows the superimposition of synthetic images over real images, providing augmented knowledge about the environment in the user's vicinity, which makes the task more pleasant and effective for the user, since the required information is spatially superimposed over real information related to it" (Danado et al., 2003, p. 1). The characteristic of augmented reality allows the expansion of the physical world information into sensory interface content with the digital transformation to perceive the virtual part that is hidden in the physical space (Mesárošová & Ferrer Hernandez, 2015). The emergence of AR applications during the industrial revolution 4.0 provides an opportunity to enhance the tourist experience by allowing them to know the unknown tourist products within the destination in an interactive and hedonic manner. In the tourism industry, the AR application has created a new and diversified experience that has changed tourists to become more interactive in their destination experience, as it allows them to explore the environment realistically even with little knowledge about unfamiliar places (Fritz et al., 2005; von der Pütten et al., 2012).

The proliferation and continuous development of the AR applications in the tourism context are seen in various tourism products such as theme parks, geological heritage sites, and historical events (Tsiotsou, 2012; Benyon et al., 2014; Martínez-Graña et al., 2013; Peter et al., 2016). The extensive information such as interactive videos, destination direction by enabling the GPS, historical information by scanning the monuments, and other tourist attractions that were integrated within the AR applications allows service providers to spread the tourism products more attractive to the global market. In the Chinese theme park, "Joy Land", for instance, AR applications enable interactions between the content and tourists, thereby enhancing the visitor experience. Due to the innovation in technologies, the theme park has become popular among tourists (Dong et al., 2011). Similarly, the world-famous Disney theme parks introduced the "Creators of the Walt Disney" to provide a novel experience by bringing old movies to life with the characters' building using AR technology (Mine et al., 2012). These highlighted examples have employed AR as a marketing tool to attract tourists by allowing them to walk through the theme park attractions more engaged and interactive, with the support of AR applications.

In 2014, the potential utilisation of the AR application also reached the urban heritage tourism to create unique visitor experiences in the recognised United Nations Educational, Scientific and Cultural Organization (UNESCO) World Heritage sites in Dublin with the young British female travellers (Tom & Jung, 2015). Tourism plays a significant role in portraying the culture and heritage of India, like many other countries. Every year, a lot of international tourists visit the famous Thanjavur Big Temple in the state of Tamil Nadu, South India, but not all get to know historical information about the temple due to the lack of a tour guide. Thus, they introduced the technology-enabled tour guide assist systems using the AR application in the temple to provide personalised tour assistance to tourists. The founder of Augtraveler, Pankaj Manchanda, developed an AR application to help everyone to learn about the heritage sites and monuments through interaction by giving importance to travellers' experience at heritage sites in India (Sindhu, 2019). Before the pandemic, Macau started to develop the mobile AR heritage application for managing heritage sites and to improve the quality of tourist experience, as it will provide updates about the crowdedness at famous sites such as Senado Square, A-Ma temple, and the Ruins of St

Paul's to avoid congestion (Yin et al., 2021). The negative impact of urban heritage tourism due to the use of space and signages with regard to the expansion of the urban destinations can be overlaid with the integrated AR applications to access all the relevant information to enhance the tourist experience (Chang et al., 1996; Jung & Han, 2014; Kalay et al., 2007; Patuelli et al., 2013). The AR applications will have the ability to provide personalised information about the destination using a market-based and interactive 3D book and videos, especially considering the current global pandemic impact to engage the people to explore and increase their intention to visit the tourist destinations in the future.

VR is defined as the system that has as its goal the complete replication of elements of the physical world with synthesised three-dimensional (3D) materials to create a computer-generated 3D environment (virtual environment), whereby the immersive experience is achieved through the use of one or more of the human five senses (Guttentag, 2010; Sussmann & Vanhegan, 2000). In the early stages of technology revolutions, the tourism industry was considered technology-friendly with the adaptation of computerised booking systems, web-based reservation, and online sale of hotel rooms. Even though the tourism industry identified the potential of VR in the year 1995 itself, the ability to immerse through the 3D plans of destination was explored only a decade back. The development phase creates a realistic and navigable virtual environment to understand and overcome the challenges involved in the execution of destination plans (Cheong, 1995; Guttentag, 2010). The VR is also used as an efficient communication tool towards the development of tourism projects by involving people from different cultures and getting their feedback via VR systems, which allow integration of multiple visual languages for people to explore the interaction and behave in a real-world setting for them to communicate effectively in future (Al-Kodmany, 2002; Guttentag, 2010; Jude & Ukekwe, 2020). In the field of sustainable tourism development, virtual reality can play an effective role by ensuring effective communication (Jude & Ukekwe, 2020). Additionally, there is a greater potential to use VR in the sales and marketing of individual businesses in the tourism industry.

The active involvement of customers in computer-generated virtual settings is imperative to create an interactive virtual reality environment and a better customer experience, as it ensures their closer attachment to the destination while exposed to VR contents with rich media features compared to static images (Kandaurova & Lee, 2019; Lee et al., 2020). The immersive experience of VR increases customer destination attachment and their intention to visit the destination with impulsive desires (Kang et al., 2020). In recent times, the importance of authentic experience has been well received, and it influenced tourists' perceived knowledge, hypothetical distance, affective intensity, and impulsive desire for a destination with the integration of VR head-mounted displays and VR applications to promote tourism (Kang, 2020; Kim et al., 2020). In the tourism industry, VR has gained attention and evolved as one of the most appreciated technological innovations for improving customer services, increasing productivity, and management to face the global competitive edge (Labanauskaitė et al., 2020; Vishwakarma et al., 2020). In certain states in Malaysia, VR is viewed as an emerging marketing tool to position their destination with interactive advertisement and promote their tourism products. Recently, the Sarawak state government emphasised promoting tourism globally with the understanding of the potential use of digital technology. The recent development and sudden shift in the focus of local companies towards VR and AR to develop contents for various attractions in Sarawak from the business perspective led to the launch of a 360-degree VR travel app in the Sarawak state to promote their destination virtually to the global market (Assiqin, 2019; Bong, 2020). Later, the development of 360-degree videos of various destinations and attractions within Malaysia is initiated through the Malaysia Travel Association as VR will enhance customers' interactive and immersive experiences in the tourism industry (Kim et al., 2020).

Use of AR/VR in the Hospitality and Tourism Industry During Crisis

The recent outbreak of the COVID-19 pandemic has severely affected the global hospitality and tourism industry with challenges of travel restrictions around the globe that have resulted in low occupancy levels and cut 50 million jobs worldwide (Pillai et al., 2021; Rogers, 2020). The unprecedented COVID-19 pandemic represents the toughest crisis that the tourism industry has had to go through to sustain the business with the revaluation, rethinking, and reshaping of the entire tourism products and services (Rogers, 2020). To deal with this crisis, the industry should embrace technological innovation to engage customers digitally through web-mediated virtual information, thereby providing vicarious experiences of destinations, garnering interest, and evoking positive emotions towards tourism products (Yung et al., 2021). Travel restrictions due to stay-home orders force customers into indirect travel experiences through advanced technology, such as VR (Rogers, 2020). VR enables customers to "visit" places around the world without physically leaving home. Customers can conveniently explore any virtual environment at any time (Guttentag, 2010). The effects of VR were investigated in various contexts in the hospitality and tourism industry, such as hotels and destinations (e.g. Kim et al., 2020; Lee et al., 2020). This pandemic has also created competition among the VR content developers to reduce their product costs and compete towards advanced VR systems for the tourism sector, focusing on human-computer interactions rather than resources on mechanics that elicit emotional responses (Pallavicini et al., 2020). With the new normal in place, tourism and hospitality businesses have had to redesign their services, products, and activities with the integration of AR and VR platforms to create an immersive experience for people staying at home and accessing the platform to reach out to the global market and to make the destination brand sustainable for future visit intention. The potential of AR and VR should be viewed as a gateway to break down the monotony of lockdown life since the proliferation of AR and VR provides the opportunity for tourism industry businesses to become resilient during this crisis time with advances in ICT to restart the tourism industry during the post-pandemic and also to unleash the potential of emerging technologies to keep the business within revival plans.

Conclusion and Recommendations

The volatility of the world requires the tourism stakeholders to embrace advanced technologies such as AR and VR to enhance their positions within the industry and to develop new e-marketing strategies to attract tourists who are staying home longer due to the restrictions of travel and social distance. In view of the recent development of AR and VR, integrated tourism products can strongly influence the customer journey with interactive engagement to reach out and inspire potential customers. The development of AR and VR experiences in tourist destinations has enhanced the perceived value of such destinations and created memorable experiences. Emerging destinations and tourism businesses should focus on a holistic approach to gaining a deeper understanding of customer expectations in the immersive experience and interaction.

Many emerging economies have adopted many AR and VR integrations in the hotel and tourism industry, while tourism destinations in emerging economies are still in the infant stages. However, the COVID-19 pandemic has changed the promotional strategies of many emerging destinations towards AR and VR integration to harmonise technology and the needs of tourists towards the survival and recovery of tourism businesses and destinations. Transformation in marketing tools will have a significant impact on all aspects of operations to enhance customer experience and attract diverse market segments. This chapter contributes to the existing literature on the evolution of AR and VR by comparing the recent and traditional dimensions of hospitality and tourism destinations. The potential of utilising AR and VR as marketing tools is discussed theoretically by highlighting the creation of immersive and interactive experiences in various tourism and hospitality products. This chapter provides practical implications for hotel managers, tour operators, destination marketers, and tourism development associations to leverage advanced technology as an emerging marketing tool to position the tourism destination virtually to trigger their intention to visit once the borders are lifted, and travel is becoming normal. The future recommendation of AR and VR in hospitality and tourism destination should focus on the rural and indigenous community whereby their income will get better to improve their quality of life to support the Sustainable Development Goals (SDG) of UNWTO.

References

Al-Kodmany, K. (2002). Visualization tools and methods in community planning: From freehand sketches to virtual reality. *Journal of Planning Literature, 17*(2), 189–211. https://doi.org/10.1177/088541202762475946

Almatourism, N. (19, 2019). Survey Paper. In G. Nikoli & A. Lazakidou (Eds.), *The impact of information and communication technology on the tourism sector.*

Aslan, A., Altinoz, B., & Özsolak, B. (2021). The nexus between economic growth, tourism development, energy consumption, and CO 2 emissions in Mediterranean countries. *Environmental Science and Pollution Research, 28*(3), 3243–3252.

Assiqin, N. (2019). *Digital tourism through virtual reality: Old Kuching Kempang heritage.* https://kuchingkampungheritage.my/2019/07/31/digital-tourism-through-virtual-reality/

Benyon, D., Quigley, A., O'keefe, B., & Riva, G. (2014). Presence and digital tourism. *AI & SOCIETY, 29*(4), 521–529.

Bong, K. (2020). Sarawak tourism takes next step in digitalization with Sarawak travel app and web-portal. https://dayakdaily.com/sarawak-tourism-takes-next-step-in-digitalisation-with-sarawak-travel-app-web-portal/

Buhalis, D. (2003). *eTourism: Information technology for strategic tourism management.* Pearson, FT/Prentice Hall.

Burns, B. L., Barney, J. B., Angus, R. W., & Herrick, H. N. (2014). Opportunity identification and stakeholder enrolment under conditions of risk and uncertainty. *Academy of Management Proceedings.* https://doi.org/10.5465/AMBPP.2014.17416

Chang, T. C., Milne, S., Fallon, D., & Pohlmann, C. (1996). Urban heritage tourism—The gobal-local nexus. *Annals of Tourism Research, 23*(2), 284–305.

Chen, C. (2020). The influence of smart tourism on tourist experience toward travel intention and satisfaction: Evidence from China. *International Journal of Marketing Studies, 12*(3). https://doi.org/10.5539/ijms.v12n3p65

Cheong, R. (1995). The virtual threat to travel and tourism. *Tourism Management, 16*(6), 417–422. https://doi.org/10.1016/0261-5177(95)00049-T

Chou, T. L., & Chanlin, L. J. (2012). Augmented reality smartphone environment orientation application: A case study of the Fu-Jen university mobile campus touring system. *Procedia e Social and Behavioral Sciences, 46*, 410–416.

Cutshall, R., Changchit, C., Pham, H., & Pham, D. (2021). Determinants of social commerce adoption: An empirical study of Vietnamese consumers. *Journal of Internet Commerce*, 1–27.

Danado, J., Dias, E., Romão, T., Correia, N., Trabuco, A., Santos, C. (2003). *Mobile augmented reality for environmental management (MARE)*. Paper presented at Eurographics 2003 Conference, Granada, Spain.

Del Chiappa, G., & Presenza, A. (2013). The use of network analysis to assess relationships among stakeholders within a tourism destination: An empirical investigation on cost Smeralda-gallura, Italy. *Tourism Analysis, 18*(1), 1–13.

Dong, D., Weng, W., Xu, W., Dong Li, Y., & Wang, L. (2011). *"Soul Hunter": A novel augmented reality application in theme parks*. Paper presented at Mixed and Augmented Reality (ISMAR), 10th IEEE International Symposium, Basel. https://doi.org/10.1109/ISMAR.2011.6143901.

Egger, R., & Neuburger, L. (2020). Augmented, virtual, and mixed reality in tourism. In Z. Xiang, M. Fuch, U. Gretzel, & W. Höpken (Eds.), *Handbook of e-tourism* (pp. 1–25). Springer.

Fritz, F., Susperregui, A., & Linaza, M. T. (2005). *Enhancing cultural tourism experiences with augmented reality technologies. The eurographics association*. Paper presented at the 6th International Symposium on Virtual Reality, Archaeology and Cultural Heritage VAST.

Garg, P., & Pahuja, S. (2020). Social media: Concept, role, categories, trends, social media and AI, impact on youth, careers, recommendations. In S. Alavi & V. Ahuja (Eds.), *Managing social media practices in the digital economy* (pp. 172–192). IGI Global.

Guttentag, D. A. (2010). Virtual reality: Applications and implications for tourism. *Tourism Management, 31*(5), 637–651. https://doi.org/10.1016/j.tourman.2009.07.003

Hua, L. Y., Ramayah, T., Ping, T. A., & Jun-Hwa, C. (2017). Social media as a tool to help select tourism destinations: The case of Malaysia. *Information Systems Management, 34*(3), 265–279.

Jude, O. C., & Ukekwe, C. (2020). Tourism and virtual reality (VR) in developing nations. *African Journal of Hospitality, Tourism and Leisure, 9*(2), 1–16.

Jung, T., Chung, N., & Leue, M. C. (2015). The determinants of recommendations to use augmented reality technologies: The case of a Korean theme park. *Tourism Management, 49*, 75–86.

Jung, T., & Han, D. (2014). Augmented reality (AR) in urban heritage tourism. *e-Review of Tourism Research*, 5, 1–5.

Kalay, Y., Kvan, T., & Affleck, J. (2007). *New heritage: New media and cultural heritage*. Routledge.

Kalmakova, A. A. (2015). Cifrovye turisticheskie jekosistemy i ih rol' v marketinge destinacij. *Geografija i Turizm: Sb.nauch.tr. 14*, 57–62.

Kandaurova, M., & Lee, S. (2019). The effects of virtual reality on charitable giving: The role of empathy, guilt, responsibility, and social exclusion. *Journal of Business Research, 100*, 571–580.

Kang, H. (2020). Impact of VR on impulsive desire for a destination. *Journal of Hospitality and Tourism Management, 42*, 244–255.

Kang, H., Shin, J., & Ponto, K. (2020). How 3D virtual reality stores can shape consumer purchase decisions: The roles of informativeness and playfulness. *Journal of Interactive Marketing, 49*, 70–85.

Kang, M., & Schuett, M. A. (2013). Determinants of sharing travel experiences in social media. *Journal of Travel & Tourism Marketing, 30*(1–2), 93–107.

Kavoura, A., & Stavrianea, A. (2014). Economic and social aspects from social media's implementation as a strategic innovative marketing tool in the tourism industry. *Procedia Economics and Finance, 14*, 303–312.

Kazandzhieva, V., & Santana, H. (2019). E-tourism: Definition, development and conceptual framework. *Tourism: An International Interdisciplinary Journal, 67*(4), 332–350.

Kim, M. J., Lee, C. K., & Jung, T. (2020). Exploring consumer behavior in virtual reality tourism using an extended stimulus-organism-response model. *Journal of Travel Research, 59*(1), 69–89.

Kononova, O., Prokudin, D., & Tupikina, E. (2020). From e-tourism to digital tourism. Terminologically review. In *CEUR workshop proceedings* (pp. 164–177). RWTH Aahen University.

Labanauskaitė, D., Fiore, M., & Stašys, R. (2020). Use of e-marketing tools as communication management in the tourism industry. *Tourism Management Perspectives, 34*. https://doi.org/10.1016/j.tmp.2020.100652

Li, J., Xu, L., Tang, L., Wang, S., & Li, L. (2018). Big data in tourism research: A literature review. *Tourism Management 68*, 301–323. https://doi.org/10.1016/j.tourman.2018.03.009

Lee, H., Guillet, D., & Law, R. (2013). An examination of the relationship between online travel agents and hotels a case study of choice hotels international and expedia.com. *Cornell Hospitality Quarterly, 54*(1), 95–107.

Lee, M., Lee, S. A., Jeong, M., & Oh, H. (2020). Quality of virtual reality and its impacts on behavioral intention. *International Journal of Hospitality Management, 90*, 102595. https://doi.org/10.1016/j.ijhm.2020.102595

Martínez-Graña, A. M., Goy, J. L., & Cimarra, C. A. (2013). A virtual tour of geological heritage: Valourising geodiversity using Google earth and QR code. *Computers & Geosciences, 61*, 83–93.

Mesárošová, A., & Ferrer Hernandez, M. (2015, October 7–9). *Art behind the mind. Exploring new art forms by implementation of electroencephalography*. Cyberworlds 2015 Conference proceedings. Gitland. Sweden.

Mine, M. R., van Baar, J., & Grundho€fer, A., Rose, D., & Yang, B. (2012). Projection-based augmented reality in Disney theme parks. *IEEE Computer, 45*(7), 32–40.

Moshnjaga, E. V. (2013). Osnovnye tendencii razvitija turizma v sovremennom mire. *Vestnik RMAT, 3*(9), 20–33.

Navío-Marco, J., Ruiz-Gómez, L.M., & Sevilla-Sevilla, C. (2018). Progress in information technology and tourism management: 30 years on T and 20 years after the internet – Revisiting Buhalis & Law's landmark study about eTourism. *Tourism Management, 69*, 460–470. https://doi.org/10.1016/j.tourman.2018.06.002.

Nedelea, A., & Bălan, A. (2010). E-tourism and tourism services consumer protection. *Amfiteatru Economic Journal, 12*(28), 492–503.

OECD Tourism Trends and Policies. (2020). https://www.oecd-ilibrary.org/sites/f528d444-en/index.html?itemId=/content/component/f528d444-en

Olmedo, M., Gomez, B., Palomares, J., & Gutierrez, J. (2018). Tourists' digital footprint in cities: Comparing big data sources. *Tourism Management, 66*, 13–25.

Pallavicini, F., Pepe, A., Ferrari, A., Garcea, G., Zanacchi, A., & Mantovani, F. (2020). What is the relationship among positive emotions, sense of presence, and ease of interaction in virtual reality systems? An on-site evaluation of a commercial virtual experience. *PRESENCE: Virtual and Augmented Reality, 27*(2), 183–201.

Patuelli, R., Mussoni, M., & Candela, G. (2013). The effects of world heritage sites on domestic tourism: A spatial interaction model for Italy. *Journal of Geographical Systems, 15*(3), 369–402.

Peter. M., Tomáś. M., Alena, M., Manuel. F. H., Branislav. K., Csaba. S., Lubomír. Š., Márío. M., Ladislav. H., Peter. B., & Radoslav. D. (2016). Use of augmented reality and gamification techniques in tourism. *e-Review of Tourism Research (eRTR), 13*(1/2), 367–381. http://ertr.tamu.edu

Pillai, S. G. K., Seo, W. S. H., & Kim, W. G. (2021). COVID-19 and hospitality 5.0: Redefining hospitality operations. *International Journal of Hospitality Management, 94*, 102869. https://doi.org/10.1016/j.ijhm.2021.102869

Pitoska, E. (2013). E-tourism: The use of internet and information and communication technologies in tourism: The case of hotel units in peripheral areas. *Tourism in Southern and Eastern Europe*, 335–344. https://ssrn.com/abstract=2289872

Rogers, S. (2020). *How virtual reality could help the travel & tourism industry in the aftermath of the coronavirus outbreak.* https://www.forbes.com/sites/solrogers/2020/03/18/virtual-reality-and-tourism-whats-already-happening-is-it-the-future/

Sarkar, S. K., & George, B. (2018). Social media technologies in the tourism industry: An analysis with special reference to their role in sustainable tourism development. *International Journal of Tourism Sciences, 18*(4), 269–278.

Shaee, S., Ghatari, A.R., Hasanzadeh, A., & Jahanyan, S. (2019). Developing a model for sustainable smart tourism destinations: A systematic review. *Tourism Management Perspectives, 31*, 287–300. https://doi.org/10.1016/j.tmp.2019.06.002

Shams, S. M. R. (2016). Branding destination image: A stakeholder causal scope analysis for internationalization of destinations. *Tourism Planning and Development., 13*(2), 140–153.

Sussmann, S., & Vanhegan, H. (2000). Virtual reality and the tourism product substitution or complement? In *Proceedings of the eighth European conference on information systems (ECIS)* (pp. 1077–1083). Wirtschaftsuniversität Wien.

Sindhu K. (2019). Retrieved on 22 May 2021, from https://yourstory.com/2019/01/experienceindia-heritage-ar-augtraveler/amp

Tom, D. M. C., & Jung, T. (2015). A theoretical model of mobile augmented reality acceptance in urban heritage tourism. *Current Issues in Tourism.* https://doi.org/10.1080/13683500.2015.1070801

Tsiotsou, R. H. (2012). Introduction to strategic marketing in tourism. In R. H. Tsiotsou & R. E. Goldsmith (Eds.), *Strategic marketing in tourism services* (pp. 147–162). Emerald.

Ustinova, M.V., & Shevchenko, M.V. (2019). Industrija gostepriimstva v jepohu cifrovizacii. *Jepoha nauki, 20*, 459–463. [In Russian].

Vishwakarma, P., Mukherjee, S., & Datta, B. (2020). Travelers' intention to adopt virtual reality: A consumer value perspective. *Journal of Destination Marketing & Management, 17*, 100456. https://doi.org/10.1016/j.jdmm.2020.100456

Von der Pütten, A. M., Klatt, J., Ten Broeke, S., McCall, R., Kraemer, N. C., Wetzel, R., Blum, L., Oppermann, L., & Klatt, J. (2012). Subjective and behavioral presence measurement and interactivity in the collaborative augmented reality game TimeWarp. *Interacting with Computers, 24*, 317–325.

Wagaw, M., & Mulugeta, F. (2018). Integration of ICT and tourism for improved promotion of tourist attractions in Ethiopia. *Appl Inform, 5*(6). https://doi.org/10.1186/s40535-018-0053-x

Wanderer. (2019). Wanderer footsteps blog (2019). Retrieved on 18 May 2021, https://wandererfootsteps.com/tourism-in-the-digital-world/

Yin, C. Z. Y., Jung, T., & tom Dieck, M. C., & Lee, M.Y. (2021). Mobile augmented reality heritage applications: Meeting the needs of heritage tourists. *Sustainability, 13*, 2523. https://doi.org/10.3390/su13052523

Yovcheva, Z., Buhalis, D., & Gatzidis, C. (2013). Engineering augmented tourism experiences. In L. Cantoni & Z. Xiang (Eds.), *Information and communication technologies in tourism 2013* (pp. 24–35). Springer.

Yung, R., Khoo-Lattimore, C., & Potter, L.E. (2021). Virtual reality and tourism marketing: Conceptualizing a framework on presence, emotion, and intention. Current Issues in Tourism, 24(11), 1505–1525, https://doi.org/10.1080/13683500.2020.1820454

Part III

Destination Marketing Process and Activities

9

New Approaches to Market Segmentation, Targeting and Positioning: The Case of Maribojoc, Bohol, Philippines

Maria Criselda G. Badilla

Introduction

Market segmentation, targeting, and positioning are key marketing concepts. Strategies for finding viable markets have been a subject matter of interest for tourism businesses and destinations. Many destinations have developed different products and services that show the beauty and character of the destinations. Tourism products are unique because they are experiences and composite products. Tourism experience is portrayed as a quest for the inversion of everyday life; the upper-middle-class tourist seeks to be a 'peasant for a day', while the lower middle class will seek to be a 'king/queen for a day' (DeCrop 2014). Broadening the scope, in view of this, Pine and Gilmore (1999) have suggested a shift from a commodity- and service-based economy to an 'experience economy' in which

M. C. G. Badilla (✉)
University of the Philippines Asian Institute of Tourism,
Quezon City, Philippines
e-mail: mgbadilla2@up.edu.ph

I. Mensah et al. (eds.), *Marketing Tourist Destinations in Emerging Economies*, Palgrave Studies of Marketing in Emerging Economies,
https://doi.org/10.1007/978-3-030-83711-2_9

the consumption experience itself becomes the primary source of value. The increasing focus on consumers as active value co-creators means that traditional segmentation goals and methodologies may no longer be relevant (Rihova et al. 2019).

A review of the literature on tourism and hospitality from 1969 to 2020 by Mulet-Forteza et al. (2021) revealed that topics related to management, with concepts such as loyalty, purchase intention, satisfaction, and repeat visit intention, have been heavily researched. In a similar review conducted by Morrison and Coca-Stefaniak (2020), there were six predominant research themes or areas of study on managing and marketing city tourism destinations, three of which are marketing related. Scholars have produced a significant number of research articles and books on specific markets, but very few have investigated the process, benefits, challenges, and issues of market segmentation, targeting, and positioning for emerging destinations.

A new shift is currently taking place, focusing on experience and offering unique ways to connect and engage with local people and gain insights into local lifestyles, cultural heritage, and sense of place (World Tourism Organization and European Travel Commission [WTO and ETC] 2017). Moreover, these experiences can include unique, more intimate, customised experiences that reflect traveller values and aspirations, or larger cultural events and festivals that reflect the personality of the territories concerned (WTO and ETC 2017), as younger tourists seek the experience more than the actual destination in itself (Fyall 2019). As tourism marketers are marketing experiences rather than mere products, identification of market segments has become increasingly challenging, as it has gone beyond usual descriptors such as country of origin, age, and income levels. Therefore, destination marketers need to have a better understanding of tourists' perspectives and be more creative in their offerings of the tourism experience. The prevalence of new trends in tourism calls for a better understanding of how the postmodern tourist markets consume touristic experiences and how emerging destinations can tap into the growing tourist market amidst the challenges they face.

Understanding the Market Through Travel Motivations

As marketing centres on identifying the needs and wants of customers, destination marketers need to identify their customers first. Identifying customers and deciding how to meet their wants and needs is a basic task facing every destination (Cook et al. 2017). Kotler et al. (2017) define a market as a set of actual and potential buyers of a product who share a particular need or want that can be satisfied through exchange relationships through the marketing process.

In emerging destinations, the challenge of finding profitable market niches needs to be overcome in order to create successful tourism destinations. Morrison (2019) argues that predominant market trends have an impact on tourism destinations. Such trends result in changes in the characteristics of tourists, the purpose of travel, trip planning, and travel arrangements. The use of technology and how tourists co-create the travel experience also impact travel motivations and destination choice. Destinations historically considered to be dependable may be less so in the future, as increasing demand for experiential tourism is driving more tourists to seek less traditional, 'venture' destinations (Plog 2001).

Understanding travel motivations and behaviours of specific markets have implications for product development ideas, targeting, positioning, and service enhancements that can attract and satisfy the specific segments. Categorising these travel motivations into segments identifies specific groups of tourists with similar needs, wants, likes, and travel behaviour patterns. This allows marketers to better understand market opportunities and maximise them. Destination marketing organisations (DMOs) of emerging markets do not feel the need to invest in market research because international markets still represent a small portion of their tourist demand, but as international demand and competition grow, research data will become more valuable for emerging destinations (Morrison 2019).

Motivations alone cannot be used for segmenting a traveller market segment. For example, youth travellers' category is limited if the subject of the investigation is too broad (Han et al. 2019b). Other variables should also be considered to fully understand the particular market. While examining a traveller segment would allow for a better understanding of the people in that particular market (Stainton 2016), geographic and demographic profiling has become quite limiting.

In fact, if destination marketers understand the 'psychographic curve' (Plog 2001), they will be able to control and maintain an ideal positioning in a destination's tourism development (Fyall 2019). The concept of experiential tourism has allowed marketers to develop psychographic approaches to customer segmentation that gives destination marketing organisations (DMOs) and operators an insight into visitors' travel values that align most closely with the destination offering (WTO and ETC 2017). This shows the importance of market segmentation, targeting, and positioning in a destination's tourism planning, development, and management process.

Market Segmentation

The task of meeting diverse tourists' needs and wants gave rise to the idea of market segmentation. Instead of meeting the needs and wants of the entire market with a single product or service, marketers divide the market into smaller but more homogeneous market segments (Cook et al. 2017). Markets with similar characteristics and needs are identified as homogeneous groups to simplify the marketing process. Marketers need to carefully select their priority target markets to make more effective use of their resources and generate the best return on investment (Morrison 2019). Beyond statistics on tourist arrivals, marketers should dig deeper into a detailed understanding of their customers' travel behaviour. Segmentation allows the sharing of similar characteristics and motivations as well as similar likes and interests (WTO and ETC 2017). Segmentation encourages an alignment with consumers' needs. It helps to find new market opportunities; predict possible consumer behaviour that may likely lead to consumption and encourage purchase; and facilitate more efficient allocation of resources through a better understanding of the market (Kotler and Keller 2012). The concept of experiential tourism has resulted in the development of psychographic approaches to customer segmentation that gives DMOs and operators an insight into visitors' travel values and those segments that align most closely with a destination's offerings (WTO and ETC 2017).

Marketing scholars provide a variety of segmentation variables which include (a) geographic, (b) demographic, (c) psychographic, and (d) behavioural (Kotler et al. 2017).

Geographic segmentation has been the simplest form of segmentation where the market is divided based on country, city, town, or neighbourhood of residence/origin. Studies profiling tourists based on country of residence are still pursued in current literature. Such studies include a study made on profiling and segmentation of international tourists in India (Singh et al. 2017). The study revealed that the majority of Asians visit to see India's rich history and culture, while Australians visit to experience spirituality and see its natural beauty. Europeans come to experience spirituality, and Americans visit to enjoy services that offer value for money. Age was also observed to attract people to India for different reasons; people aged 19–30 visit India for its spirituality, natural beauty, and food; and people aged 31–50 are attracted due to diverse factors, while visitors 51 years old and above visit India for spiritual activities (Singh et al. 2017).

Demographic profile looks into segmenting the market based on age, gender, income, occupation, education, religion, race, and life cycle (Kotler et al. 2017). Studies on the use of demographic profiling are not uncommon in the literature. Some examples of these studies include a segmentation analysis of cultural heritage tourism through the lens of youth in Thailand (Pansukkum and Swanson 2018) and how food value video clips were effective in promoting food tourism for Chinese Generation Y versus non-generation Y (Xu and Pratt 2018). Also, studies on religious tourism have recently been published as the Muslim tourism market is rising in significance, such as developing a Muslim tourism market in Taiwan (Hsu et al. 2020), and Halal-friendly hotels in Thailand (Jeaheng et al. 2019).

Psychographic segmentation involves grouping people on how they live, their priorities, and their interests; describing a person's lifestyle and personality; and showing similarities in hobbies, sports preferences, and musical interests (Cook et al. 2017). This type of segmentation is based on social class, lifestyle, and personality (Kotler et al. 2017). The growth of niche markets was mainly due to the growing trend of psychographic segmentation. Adventure, cultural, food, leisure, farm, business, wellness, eco-tourism, and many others are examples of psychographic segmentation. The farm tourism segment in Taiwan (Chen et al. 2010) reveals the lifestyle characteristics of farm tourists.

Behavioural segmentation refers to dividing groups based on their knowledge, attitude, use of or response to a product or service (Badilla 2015). These behavioural variables include special occasion segmentation, benefits sought, user status, user rate, loyalty status, and buyer readiness. Scholars have examined behavioural segmentation in studies such as examining honeymooners' tourist behaviour in Thailand (Fakfare et al. 2020).

Product-related segmentation groups are travellers who seek different benefits based on their reason for travel together (Cook et al. 2017). The most popular reasons for travel are leisure, business, and visiting friends and relatives. Destinations need to consider developing products that will focus on specific segments that have the highest interest levels in using the specific products (Morrison 2019). Product-related segmentation is key to tapping the Muslim market, whereby marketers need to understand the special needs dictated by one's religion (Hsu et al. 2020).

There seem to be some overlaps in market segmentation using geodemographic segmentation, socio-economic profiles, and psychographic and behavioural. A variety of market segments have been identified, such as geodemographic on the Chinese Generation Y Market (Xu and Pratt 2018), using both geographic (country) and age (generation y) as variables for segmentation. Another example is that volunteer tourism and youth travellers as an emerging trend (Han et al., 2019) combine psychographic (volunteerism) and demographic (youth) variables.

New strategies for segmentation have emerged in the constantly changing travel trends. The literature identifies a variety of segmentation strategies used by destinations veering away from the one-size-fits-all marketing approach. These approaches include benefit segmentation, practice-based approach, stakeholder approach, and explorer quotient.

Benefit segmentation was first introduced by Haley (1968) as a segmentation approach that looks into how the benefits sought by consumers determine their purchase behaviour more accurately than merely descriptive or demographic characteristics. Based on a study conducted by Nduna and vanZyl (2020) of Mpumalanga Province, South Africa, a benefit segmentation framework was proposed for a nature-based tourism destination using cluster analysis. Segmenting tourists using benefits sought identifies benefits relevant to consumers' needs and wants in order to design products and promotional messages to specific market segments identified.

Practice-based segmentation looks into social practices as an alternative way of segmenting the market. With the emerging world view of tourists as co-creators of the tourism experience, Rihova et al. (2019) conducted exploratory and evaluator research on practice-based segmentation as a potential conceptual alternative to more traditional segmentation approaches. They argue that social practices represent an altogether different starting point for strategic, customer-induced segmentation. Further research on co-creation practices as a basis for segmentation is highly recommended.

The *stakeholder approach* to destination segmentation takes the view of tourism stakeholders through a two-step process. The first involves understanding how multiple destination stakeholders view their market, and the second is segmenting the destination based on variables that were commonly identified by a broad range of destination stakeholders (Tkaczynski and Rundle-Thiele 2011). Using the four recognised bases of segmentation (geographic, demographic, psychographic, and behavioural), stakeholders classify their tourists based on the use of services offered to tourists, namely accommodations, mode of transportation, dining, social interactions, and holiday patterns.

Other market segmentation strategies used by developed countries can be used to segment the markets for emerging economies. The *explorer quotient* has been developed by Destination Canada, which goes beyond traditional market research of defining people but looks into an individual's personal beliefs, social values, and views of the world to learn exactly why different types of travellers seek out entirely different types of travel experiences (Destination Canada 2015). Data-driven market segmentation is proposed by academic and industry data analysts, which provides a better quality of market segmentation solution derived from empirical tourist data which can translate into a competitive advantage and increased business for destinations (Hajibaba et al. 2019). Market research studies on emerging destinations have been scarce.

Segmentation permits tourism service providers to better meet specific customer needs and wants while attempting to increase their satisfaction (Cook et al. 2017). Destination marketing organisations carefully select priority target markets to ensure that they make more effective use of their resources and generate the best return on investment (Morrison 2019). Destination marketers are increasingly streamlining their marketing efforts towards identified segments rather than a one-size-fits-all approach (Nduna and vanZyl 2020). Segmentation provides a tool for destinations to effectively plan their marketing efforts (Nduna and van-Zyl 2020). Destinations need to identify the personal motivations of those who visit and those who are interested in the destination, determining what draws them to the destination, what is preventing them from visiting, and what their needs/wants are from the destination in question (Fyall 2019) to have a greater understanding of personality profiles of the destinations' target markets. Beyond statistics, marketers need to know why tourist behaviours are changing the way they do (Morrison 2019).

Specialised Tourist Segments

Among the challenges of market segmentation are (a) some markets are too small to segment, (b) some markets not considered a few years ago are growing in size and significance such as travellers with disabilities and volunteerism, and (c) individuals could be classified as members of many different markets (Cook et al. 2017). Further, it has also become difficult to identify emerging markets in international tourism, and competition among destinations has intensified greatly (Morrison 2019). Because of the size and significance of specific market segments, some have developed smaller segments within the larger market segments.

For instance, food consumption has become an important part of the tourism experience. This gave rise to the term 'foodies', which is defined as people with a passion for eating and learning about food but are actually not food professionals (Cairns et al. 2010). Such foodie travellers who seek out unique food experiences have different preferences and behaviours. In a study conducted in North California, USA, three sub-segments of foodies were identified, and these were known as enthusiasts, creators, and samplers (Kline et al. 2018). Compared to another study of Chinese foodies who have travelled to Taiwan, correlations existed among foodies' consumption motivations, emotions, perceived well-being, and experiential values (Chang et al. 2021). The findings revealed that foodies give a high premium to food experience; hence, preparers of food should emphasise local cultural characteristics to make the food consumption experience unique. In Ecuador, a study showed the existence of three types of tourists based on their interest in local gastronomy during travel. Based on the interest and previous cultural motivation of tourists, the typologies included survivors, enjoyers, and experiencers (Galvez et al. 2020).

Further, the volunteer tourism industry started as a micro-segment but has now developed a variety of smaller segments, which include medical, environmental, and cultural niches (Stainton 2016). A study conducted with 342 highly educated youths from 58 countries came up with a structural model assessment of the volunteer travel motivations, which include altruism, personal development, education, ego-enhancement, and escapism (Han et al., 2019). By identifying these travel motivations, marketers will be able to target and attract these specialised tourist segments such as the foodies and volunteer tourists by designing communication strategies that are congruent with their travel motives.

Another example of niche segmentation, wine tourists, has varied typologies. In India, wine-related lifestyle was segmented in the context of urban Indian consumers. The study yielded six kinds of wine-related lifestyles, namely drinking ritual, social drinking, consumption reason (mood, enjoyment, and relaxation), consumption practice, consumption planning, and quality (Sharma et al. 2020). Further, through clustering, three typologies were identified as cautious social drinkers, lone regular drinkers, and highly engaged drinkers (Sharma et al. 2020). If a niche market as small as wine tourists can be segmented into different typologies, there are boundless opportunities for research on market segmentation of other types of markets.

In order for these specialised markets to be accessible to marketers, these are also described based on countries of origin, age or generational groups, gender, and level of education, among others.

Market Targeting

Market targeting involves evaluating each segment's attractiveness and selecting the ones on which the business can operate (Kotler et al. 2017). Kotler et al. (2017) identified three factors used for evaluating market segments, namely segment size, attractiveness, and company's objectives and resources. Segment size refers to sales growth and potential for profit. Segment attractiveness shows the profitability of the segment in relation to the presence of competitors, market saturation, and demand. The company's objectives should be aligned with the markets being identified, and the company has enough resources to succeed in the specific market segment.

There are several ways of tapping market segments based on a company's resources. These are the degree of product homogeneity, market homogeneity, and competitors' strategy. Kotler et al. (2017) recommend market coverage strategies as follows: (a) *undifferentiated marketing* where a company ignores segmentation and goes after the entire market with one offer, (b) *differentiated marketing* where the company targets several market segments with different product offers to capture a bigger chunk of the market, (c) *concentrated marketing* where a company allocates most of its resources to a specific market segment to pursue a substantial market share of a few smaller markets, and (d) *micromarketing* where a company tailors its products and marketing programmes to the tastes of specific individuals and locations. The market segments can be tapped through an efficient marketing communications plan to create awareness, interest, and desire for the tourism products and services offered by destinations.

The use of social media (SM) platforms has become more prevalent and is the preferred channel to convey messages to specific target markets (Yetimoglu and Ugurlu 2020). Instagram, Facebook, Twitter, and YouTube are the most commonly used channels in effectively tapping target markets. Social media is a cost-effective digital platform for attracting potential customers and promoting tourism products and services through direct interaction with customers (Alalwan et al. 2017; Lin et al. 2020; Styvén and Wallstrom 2017).

Tourism marketers have seen the growing importance of social media influencers in conveying information and messages to their followers, as these are perceived to be with higher credibility than destination-generated content (Yetimoglu and Ugurlu 2020).

A Muslim market is a new market segment that has become very attractive for emerging economies. There are over two billion Muslims globally (worldpopulationreview.com), and interest in tourism has been growing. Taiwan wants to diversify its inbound tourist market by attracting Muslims from Southeast Asia, India, Middle East, and China (Hsu et al. 2020). Marketers, however, need to consider the special needs of this market segment. In order for emerging destinations to tap into this market, they need to comprehend that their travel requirements are dictated by Islamic rules and obligations, which include Halal food, toilets, prayer facilities, transportation, and entertainment (Hsu et al. 2020).

Market Positioning

Positioning is a natural extension of market segmentation and targeting, with destinations constantly under review with regard to a myriad of factors such as destination attributes, price, competitor destinations, and type of tourist (Chacko and Marcell 2008).

Once destinations have identified their target markets, they need to identify the appropriate value proposition for differentiation and positioning (Kotler et al. 2017). Market positioning involves developing a competitive market position for the product and designing a marketing mix to achieve and maintain the desired position (Kotler et al. 2017). Several positioning strategies based on specific product attributes, service offered, personnel quality, location, and brand image can be used (Kotler et al. 2017) to help identify its competitive advantage and unique selling proposition.

DMOs play an important role in developing the positioning and branding of destinations. In the cities of Delhi, Mumbai, and Kolkata, food tourism is emphasised in their place branding initiatives but has been found to fall short in positioning themselves as food urban destinations (Amore and Roy 2020). Enhancing the visibility of food trails by interacting with international tourists on different social media platforms and the quest for authenticity in the urban visitor experience were some of the recommendations to enhance India's positioning as an urban food tourism destination (Amore and Roy 2020). On the other hand, positioning a destination may also vary depending on who is marketing it. A study on Islamic marketing proposed a framework on how to position a destination to the Muslim market, namely using the halal certification for non-Muslim marketers and Islamisation for Muslim marketers (Islam 2020).

Defining one's positioning successfully is key to creating a hierarchy of visitor experiences within the destination, ranging from 'iconic', 'signature', or 'must see' experiences and attractions, to supporting experiences, such as local foods and gastronomy (WTO and ETC 2017). With the emergence of new destinations around the world, the marketplace has become more competitive. Destinations need to maintain or increase their competitive position to be truly distinctive in the marketplace, as destinations that are historically dependable may be less so in the future because increasing demand for experiential tourism is driving tourists to newer destinations (Fyall 2019).

Role of Technology in Market Segmentation, Targeting, and Positioning

Information and communications technologies (ICTs) provide opportunities for market segmentation and market research. Despite an abundance of academic papers on market segmentation in tourism, there are only a few focusing on information and communications technology (Pesonen 2013). Technology has increased the capacity of marketers to analyse voluminous data through computer software. Segmentation has become more sophisticated with the advancements in computer hardware and software (Bowen 1998). The use of statistical data modelling software provides easy identification of market segments and can compare the effectiveness of using different software techniques in identifying market segments. Bowen (1998) argues that with computerised methods, the steps to market segmentation are combined such that computer software can give information on whom to target for specific products, hence making the process easier for marketers. Through the use of digital techniques, segmentation by computers has become bias-free, absent of the bias of human judgement. With innovations such as data mining and big data analytics providing processed information, segmentation methods have prospered.

Also, advances in digital and mobile technology have given rise to location-based marketing. As consumers become more dependent on smartphones with GPS capabilities, companies can engage consumers with real-time and location-based marketing offers (Kotler et al. 2017). SoLoMo (social + local + mobile) is a new innovation that reaches on-the-go digital tourists as they navigate destinations. In recent years, targeting specific market segments is no longer a challenge for tourism marketers. Social media networks have drawn millions to seek travel information online, and tourism marketers have heavily relied and engaged on social media as an information and promotional channel (Morrison 2019).

The role of social media has grown in significance. As consumers travel more, they have become more dependent on social media for information and no longer rely on travel agents and tour operators (Wee 2017). The tourism and hospitality literature reveals a growing interest in the adoption and use of social media for management and other business activities, particularly marketing (Lin et al. 2020). This has also given marketers the ability to engage customers one on one, where marketing messages have become more personalised. Social media has allowed marketers to veer away from the one-to-many marketing model that has been prevalent in the past. Businesses have been able to communicate directly to customers, while consumers have also been able to directly communicate back to businesses and other consumers. This has allowed new and emerging destinations, which do not have as much marketing budgets as more established and popular destinations to tap into the market previously dominated by such destinations.

The use of ICT in tourism is expected to grow in the coming years. With COVID-19 boosting the use of online technology for several purposes globally, such as the use of robots, artificial intelligence, and service automation, digital media provide destinations with the opportunity to be in the online world. However, emerging economies where ICT infrastructure is not fully developed will continue to experience the digital divide. Nevertheless, with online, word of mouth, and the tourist's active role in the co-creation of destination image and the tourist experience, new and emerging destinations can flourish. Technology convergence will ultimately lead to convergence of digital and traditional forms of marketing, which is seen to be more personalised and more customer-engaged (Kotler et al. 2016).

Steps in Market Segmentation, Targeting, and Positioning

Market segmentation, targeting, and positioning are valuable tools for the marketing planning process. When segmenting a market, it is crucial for marketers to identify bases for segmenting the market and develop tourist profiles for the resulting segments. Targeting the market segment entails developing measures of segment attractiveness and selecting the markets to target. In market positioning, marketers should develop positioning strategies for each target segment and develop a marketing mix for each target segment. Each step in the process is identified in Fig. 9.1.

Traditionally, these steps happen one after another in a sequential manner. However, Bowen (1998) argues that these steps to market segmentation, targeting, and positioning are no longer sequential, especially with the rise of technology that has allowed the merging of segmentation and targeting through sophisticated data analytical tools.

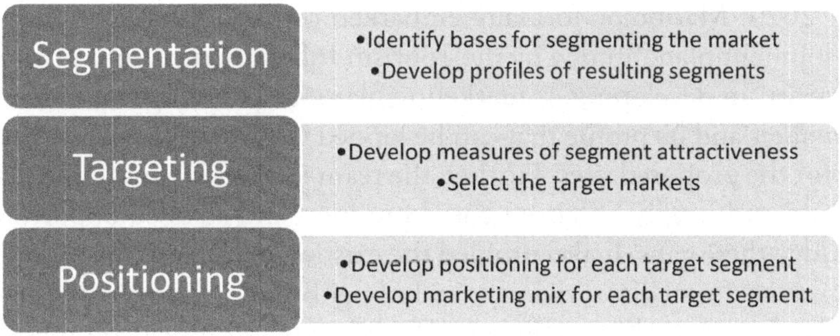

Fig. 9.1 Steps in the process of market segmentation, targeting, and positioning. (Source: Kotler et al. (2008))

The Maribojoc, Bohol, Philippines Experience

The Municipality of Maribojoc is a fourth-class municipality in Bohol, Philippines. It is divided into 22 *barangays* (Filipino term for the smallest unit of government, villages), with a total land area of 6956.8 hectares. It is a coastal town located in the southwestern portion of Bohol. Its famous attractions include the Punta Cruz Historical Watchtower, the centuries-old Holy Cross Church made of coral stones, the scenic Maribojoc Bay, the panoramic mountain ranges, the emerging Abatan River Tour, and its hospitable and peace-loving people (Bohol Philippines Travel Guide Tours and Packages n.d.). It is endowed with clear rivers, springs, and creeks that provide potable water for the people and irrigation for its rice fields. The rivers have also been adequate for swimming and river adventure tours and sports activities such as paddle boating and kayaking. It is also considered a high-tourism-value municipality in terms of the presence of heritage buildings and other interesting features that have the potential to attract tourists. The local government of Maribojoc had a strong desire to maximise the use of its resources for its tourism potential. With the popularity of Panglao Island and the capital city of Tagbilaran, it paved the way for the expansion of tourism development to the other towns of Bohol, including Maribojoc, which is 20 min (13.3 km) away from Tagbilaran and one h away (31.9 km) from Panglao Island, where the new Panglao International Airport is located.

In 2020, Maribojoc formally embarked on the creation of a tourism development plan, funded by the Tourism Infrastructure Enterprise Zone Authority. In developing its marketing plan, the planning team identified the market and its profile that can be tapped to give value to the planning area for the proposed sites. Further, the team identified gaps in the market demand and supply of leisure-based products and assessed opportunities to bridge these gaps. It also profiled the existing markets of the municipality and its immediate vicinity, including Bohol's flagship destination, Panglao Island, and its capital city, Tagbilaran, and provided other destinations as a benchmark for the destination development of Maribojoc.

Geographically, the top 10 international markets of Bohol are China, Korea, Taiwan, USA, France, Germany, Japan, Australia, United Kingdom and Canada (DOT Region 7 report). In 2018, 43% of tourists are non-Philippine residents and 57% are Filipinos with an average length of stay of 5-6 nights. The main reasons for travel is leisure and recreation followed by business travellers and the tourists' preferred travel companions were their family and friends.

Initially, three major tourism attractions were identified as the focus of development (Table 9.1). These three attractions are Punta Cruz watchtower, Anislag River complex covering the several barangays and Padhan Prayer Mountain. One of the constraints of emerging destinations is its lack of funding thereby not allowing for the conduct of a more methodical market research other than stakeholder consultations, analysis of prevailing tourist trends and benchmarking with existing profile of tourists to Bohol.

Table 9.1 Proposed product development themes and target markets

		Proposed target markets		
Sites	Proposed development	Primary	Secondary	Tertiary
Punta Cruz	Eco-cultural development	Weddings and events	Leisure travellers	Educational/ training
Padhan Prayer Mountain	Pilgrimage/ religious	Religious pilgrims	Nature/ adventure seekers	Cultural
JANPPOBA River Network	River cruise	Leisure tourists	Nature	Cultural

The *destination weddings market* is a growing trend in the Philippines. Filipino couples tend to have longer engagement periods (12–24 months) to be able to plan their wedding day better. They choose a venue that has a romantic atmosphere and historical ambience and are excited to go out of town to celebrate in popular wedding destinations in the Philippines. With Punta Cruz's historical significance, it is envisioned as a premier events place in Bohol. Destination weddings now comprise 29.4% of the wedding market in the Philippines (bridestory.com). The *religious market* is known for its hallmark pilgrimage sites of the Holy Land and Mecca. Pilgrimage is a well-known religious phenomenon and exists in all the religions in the world, including Buddhism, Hinduism, Islam, Judaism, and Christianity (Collins-Kreiner and Kliot 2000). Maribojoc's centuries-old church and Padhan Prayer Mountain look into attracting pilgrims from all over the country. One of the most important travel motivations is leisure. Dann (1977) and Crompton (1979) identified push factors of travel motivation, which include escape from a perceived mundane environment, exploration and evaluation of self, relaxation, prestige and facilitation of social interactions as major travel motivations. Most of these motivations point to the leisure travel market.

The JANPPOBA river network seeks to attract the leisure market, with its river cruise, floating restaurants, and immersive activities, complementing existing eco-tours such as the firefly watching tour at night, nipa weaving, mangrove planting, and coconut climbing. In the UNWTO Highlights 2018, 55% of tourists travel for the main purpose of leisure, recreation, and holidays. Special interest tourism niches will also be tapped for Maribojoc, and these include the eco-adventure market, culture/heritage tourist, and the educational tourist.

The product development happens alongside the target market identification. Designing, communicating, and selling a product or service is difficult if it does not match consumers' expectations (DeCrop 2014). A product-market fit is crucial to the success of tourism marketing in emerging destinations. Maribojoc is positioning itself as the premier eco-cultural destination in Bohol, combining its environmental assets and the culture and heritage of the town to make it a worthwhile destination for tourists to extend their stay in Bohol.

Conclusion

The postmodern tourist is no longer satisfied with a passive role as a consumer (DeCrop 2014). Identifying tourism market segments and deciding how to meet their wants and needs are basic tasks that tourism marketers are faced with (Cook et al. 2017). As tourists wish to become co-creators of value and co-producers of travel activities, creating value now lies in both the hands of tourism providers and tourists (DeCrop 2014). Emerging destinations face the challenge of providing tourism experiences while maximising the tourist's role as a co-creator of experiences.

As statistics on tourist arrivals are not enough to identify the customer base, market research and market-planning techniques can provide a more thorough understanding of specific markets (Morrison 2019). Market segmentation continues to play a vital role in tourism market planning. The rise of sophisticated ICT tools for market segmentation allows emerging destinations to easily identify market segments to target without very costly market research programmes. With the emergence of many new destinations around the world, perhaps the most challenging task is the ability of destinations to be truly distinctive in the marketplace (Fyall 2019). Digital marketing and creativity can provide smaller destinations without huge marketing budgets the opportunity to be noticed in a crowded marketplace. As such, tourism marketers need to be updated on the latest trends in the market to stay relevant and competitive in the marketplace. Understanding tourists is more critical today, as they are more demanding and changeable, showing more complex preferences and decision patterns (DeCrop 2014). Emerging destinations that are able to respond to market trends, opportunities, and challenges can become successful in creating niches for themselves.

Acknowledgements The author acknowledges Mayor Romulo Manuta and Tourism Officer Mary Annie Jabines of the town of Maribojoc for granting permission to use the information about the town for this chapter.

References

Alalwan, A., Rana, N., Dwivedi, Y., & Algharabat, R. (2017). Social media in marketing: A review and analysis of the existing literature. *Telematics and Informatics, 34*(7). https://doi.org/10.1016/j.tele.2017.05.008

Amore, A., & Roy, H. (2020). Blending foodscapes and urban touristscapes: International tourism and city marketing in Indian cities. *International Journal of Tourism Cities, 6*(3), 639–655.

Badilla, M. C. (2015). *Tourism marketing*. Rex Bookstore.

Bohol Philippines Travel Guide Tours and Packages. (n.d.). https://www.bohol-philippines.com/

Bowen, J. (1998). Market segmentation in hospitality research, no longer a sequential process. *International Journal of Contemporary Hospitality Management, 10*(7), 289–296.

Cairns, K., Johnston, J., & Baumann, S. (2010). Caring about food doing gender in the foodie kitchen. *Gender & Society, 24*(5), 591–615.

Chacko, H. E., & Marcell, M. H. (2008). Repositioning a tourism destination: The case of New Orleans after hurricane Katrina. *Journal of Travel & Tourism Marketing, 23*(2–4), 223–235.

Chang, J., Morrison, A., Lin, S., & Ching-Yen, H. (2021). How do food consumption motivations and emotions affect the experiential values and well-being of foodies? *British Food Journal, 123*(2), 627–648.

Chen, J., Chang, L., & Cheng, J. (2010). Exploring the market segments of farm tourism in Taiwan. *Journal of Hospitality Marketing, 19*, 309–325.

Collins-Kreiner, N., & Kliot, N. (2000). Pilgrimage tourism in the holy land: The behavioural characteristics of Chirstian pilgrims. *GeoJournal, 50*, 55–67.

Cook, R., Hsu, C., & Taylor, L. (2017). *Tourism: The business of hospitality and travel* (6th ed.). Pearson.

Crompton, J. L. (1979). Why people go on pleasure vacation. *Annals of Tourism Research, 6*(4), 408–424.

Dann, G. (1977). Anomie, ego-enhancement and tourism. *Annals of Tourism Research, 4*, 184–194.

DeCrop, A. (2014). Theorising tourist behavior. In S. McCabe (Ed.), *The Routledge handbook of tourism marketing* (pp. 251–267). Routledge.

Destination Canada. (2015). Ambitious together: 2015 annual report. https://www.destinationcanada.com/sites/default/files/201611/Corporate_Annual Report_2015_EN.pdf

Fakfare, P., Lee, J., & Ryu, K. (2020). Examining honeymoon tourist behavior: Multidimensional quality, fantasy and destination relational value. *Journal of Travel and Tourism Marketing, 37*(7), 836–853.

Fyall, A. (2019). Tourism destination re-positioning and strategies. In E. Fayos-Sola & C. Cooper (Eds.), *The future of tourism* (pp. 271–283). Springer.

Galvez, J., Torres-Matovelle, P., Molina, G., & Santa Cruz, F. (2020). Gastronomic clusters in an Ecuadorian tourist destination: The case of the province of M Manabi. *British Food Journal, 122*(12), 3917–3934.

Hajibaba, H., Grun, B., & Dolcinar, S. (2019). Improving stability of market segmentation analysis. *International Journal of Contemporary Hospitality Management*. https://doi.org/10.1108/IJCHM-02-2019-0137

Haley, R. (1968). Benefit segmentation: A decision-oriented research tool. *Journal of Marketing, 32*(3), 30–35.

Han, H., Al-Ansi, A., Koseoglu, M., Lin, P., Park, J., Yu, J., & Kim, W. (2019a). Halal tourism: Travel motivators and customer retention. *Journal of Travel and Tourism Marketing, 36*(9), 1012–1024.

Han, H., Meng, B., Chua, B., Ryu, H., & Kim, W. (2019b). International volunteer tourism and youth travelers: An emerging tourism trend. *Journal of Travel and Tourism Marketing, 5*(36), 549–562.

Hsu, P., Ku, E., Lai, T., & Hsu, S. (2020). Developing a Muslim tourism market: The perspective of travel agencies. *Journal of Hospitality and Tourism Insights*. https://doi.org/10.1108/JHTI-08-2020-0155

Islam, M. (2020). Segmenting, targeting and positioning in Islamic marketing. *Journal of Islamic Marketing*. https://doi.org/10.1108/JIMA-10-2018-0181

Jeaheng, Y., Al-Ansi, A., & Han, H. (2019). Halal-friendly hotels: Impact of halal-friendly attributes on guest purchase behaviors in the Thailand hotel industry. *Journal of Travel & Tourism Marketing, 36*(7), 729–746.

Kline, C., Seungwoo, J. L., & Knollenberg, W. (2018). Segmenting foodies for a foodie destination. *Journal of Travel and Tourism Marketing, 35*(9), 1234–1245.

Kotler, P., Armstrong, G., Wong, V., & Saunders, J. (2008). *Principles of marketing*. Prentice Hall.

Kotler, P., Bowens, J., Makens, J., & Baloglu, S. (2017). *Marketing for hospitality and tourism* (7th ed.). Pearson.

Kotler, P., Kartajaya, H., & Setiawan, I. (2016). *Marketing 4.0.: Moving from traditional to digital*. John Wiley & Sons.

Kotler, P., & Keller, K. (2012). *Marketing management* (14th ed.). Pearson Education.

Lin, H., Lyu, T., Ho, W., Xu, Y., Hsieh, T., & Zhang, L. (2020). Task-technology fit analysis of social media use for marketing in the tourism and hospitality industry: A systematic literature review. *International Journal of Contemporary Hospitality Management, 32*(8), 2677–2715.

Morrison, A. (2019). *Marketing and managing tourism destinations* (2nd ed.). Routledge.

Morrison, A., & Coca-Stefaniak, J. (2020). *Routledge handbook of tourism cities*. Routledge.

Mulet-Forteza, C., Lunn, E., Merigo, J., & Horrach, P. (2021). Research progress in tourism, leisure and hospitality in Europe (1969-2018). *International Journal of Contemporary Hospitality Management, 33*(1), 48–74.

Nduna, L., & vanZyl, C. (2020). A benefit segmentation framework for a nature-based tourism destination: The case of Kruger, panorama and Lowveld areas in Mpumalanga Province. *International Journal of Tourism Cities, 6*(4), 953–973.

Pansukkum, & Swanson. (2018). Cultural heritage tourism through the lens of the youth: The case of Thai youth visitors to Ayutthaya Historical Park, Thailand. In Y. Wang, A. Shakeela, A. Kwek, & C. Khoo-Lattimore (Eds.), *Managing Asian destinations* (pp. 261–277). Springer.

Pesonen, J. A. (2013). Information and communications technology and market segmentation in tourism: A review. *Tourism Review, 68*(2), 14–30.

Pine, B., & Gilmore, J. (1999). *The experience economy: Work is theatre and every business is a stage*. Harvard Business Press.

Plog, S. (2001). Why destination areas rise and fall in popularity. *Cornell Hotel and Restaurant Administration Quarterly, 42*(3), 13–24.

Rihova, I., Moital, M., Buhalis, D., & Gouthro, M. (2019). Practice-based segmentation: Taxonomy of C2C co-creation practice segments. *International Journal of Contemporary Hospitality Management, 31*(9), 3799–3818.

Sharma, P., Singh, S., & Misra, R. (2020). Wine-related lifestyle segmentation in the context of urban Indian consumers. *International Journal of Wine Business Research, 32*(4), 503–522.

Singh, N., Krentler, K., & Ahuja, S. (2017). Profiling and segmentation of international tourists in India. *Anatolia, 28*(1), 31–40.

Stainton, H. (2016). A segmented volunteer tourism industry. *Annals of Tourism Research, 61*, 213–267.

Styvén, M., & Wallstrom, A. (2017). Benefits and barriers for the use of digital channels among small tourism companies. *Scandinavian Journal of Hospitality and Tourism, 19*(1), 1–20.

Tkaczynski, A., & Rundle-Thiele, S. (2011). Segmenting destinations: In the eyes of the stakeholders. *International Journal of Culture, Tourism and Hospitality Research, 5*(3), 255–268.

Wee, C. (2017). Emerging trends in tourism: Opportunities, challenges and implications. *Journal of Tourism and Hospitality, 6*(6), 1–5.

World Tourism Organization and European Travel Commission. (2017). Targeting the customer. *Handbook on marketing transnational tourism themes and routes*. UNWTO.

Xu, X., & Pratt, S. (2018). Social media influencers as endorsers to promote travel destinations: An application of self-congruence theory to the Chinese generation Y. *Journal of Travel and Tourism Marketing, 35*(7), 958–972.

Yetimoglu, S., & Ugurlu, K. (2020). Influencer marketing for tourism and hospitality. In A. Hassan & A. Sharma (Eds.), *The emerald handbook of ICT in tourism and hospitality* (pp. 131–148). Emerald.

10

Homecoming Events and Diaspora Tourism Promotion in Emerging Economies: The Case of the Year of Return 2019 Campaign in Ghana

Ishmael Mensah

Introduction

Special events have become an established feature of contemporary destination marketing, and destination marketers hardly miss out on opportunities to capitalise on notable events or dates (Getz, 1997). Destinations usually employ special events as a strategy for attracting tourists, including their diaspora. Events targeting the diaspora often incorporate diasporic sentiments and emotions to appeal to the diaspora and serve as catalysts for long-term destination marketing activities (Morgan et al., 2003). For instance, West African countries including Ghana have capitalised on their slavery heritage to stage events relating to the transatlantic slave trade specifically to appeal to the African Diaspora. Through such events, diaspora communities gain a sense of belonging by reconnecting

I. Mensah (✉)
University of Cape Coast, Cape Coast, Ghana
e-mail: ikmensah@ucc.edu.gh

I. Mensah et al. (eds.), *Marketing Tourist Destinations in Emerging Economies*, Palgrave Studies of Marketing in Emerging Economies,
https://doi.org/10.1007/978-3-030-83711-2_10

with their roots, their ancestral homeland and its people (Basu, 2005). This phenomenon has also been described variously as diaspora tourism, roots tourism, genealogical tourism and homeland trip.

According to Morgan et al. (2003), in contemporary times, diaspora tourism or the ancestral ties market has become an important market niche for many destinations that continue to aggressively pursue this market. This includes staging homecoming events as exemplified by Scotland and Wales (Sim & Leith, 2013). Homecoming Scotland 2009 (HS09), was a ten-month programme involving over 400 events, which aimed at attracting visitors of Scottish birth, descent, or people with an affinity to Scotland. Also, Wales Tourist Board's (WTB) 'Homecoming 2000—Hiraeth 2000' initiative stimulated and marketed a programme of events to the Welsh diaspora and international audiences with an affinity to Wales. Attracting the diaspora of a country is particularly imperative because apart from strengthening social ties and family bonds, visits by the diaspora can rekindle their commitment to their home country, and this could result in social, political, or economic investments (Asiedu, 2005).

In Ghana, the diaspora market comprises two categories of people: (a) Africans who have been dispersed on other continents through the transatlantic slave trade and (b) contemporary Ghanaian migrants living in other foreign countries (Ankomah et al., 2012; Kleist, 2013). However, within the diaspora market, Ghana's niche market is the African American market. Ghana's numerous heritage attractions connected with the transatlantic slave trade and its relatively stable political atmosphere provide good prospects for positioning the country on the African American market (Mensah & Amissah, 2009). Since independence in 1957, successive governments have continued to engage with the African American market niche through events, heritage programmes, and other marketing programmes. However, in 2018, the government of Ghana launched a year-long series of events dubbed 'the Year of Return 2019' (YOR19) to woo the African diaspora market. Ghana capitalised on its long-standing history with the African diaspora market, especially African Americans dating back to the independence era. When Ghana gained independence in 1957, it became the territorial base for the pan-African movement under the leadership of President Kwame Nkrumah. This resulted in a number of high-profile African Americans, including W.E.B. Du Bois, relocating to Ghana. There had been previous attempts at attracting the African Diaspora through events such as the Pan-African Historical Theatre Festival (PANAFEST) and Emancipation Day, as well as homecoming campaigns like the Joseph Project with varying results. The year-long YOR19 initiative sought to capitalise on the historical links between Ghana and the African Diaspora to boost Ghana's tourism industry by positioning Ghana as a key travel destination for African Americans and the African Diaspora.

The tourist destination has been widely considered to be one of the most difficult products to manage and market due to the fact that it comprises numerous products, stakeholders, organisations, and individuals that combine to deliver the destination 'product' (Fyall et al., 2006). More so, in the context of a developing African country where lack of resources and significant financial, cultural, and infrastructural obstacles also hamper efforts aimed at promoting tourism (Poirier, 2000), the issue is even more complicated when the destination is promoting a roots tourism product as in the case of Ghana. MacGonagle (2006) asserts that the promotion of roots tourism is a delicate issue. Promoting roots tourism presents a challenge to most governments because it is a form of tourism that is fraught with complexities, contradictions, and challenges (Mensah, 2021). There were some concerns that promoting year-long roots tourism events could present a tall order. Thus, there is the need to evaluate the YOR19 campaign to determine whether the desired impacts were achieved. Ritchie and Crouch (2003) argue that the evaluation of destination success must be addressed through an assessment of both input (e.g., effective marking) and output variables (e.g., increase in visitation). Using the year of return as a case study, this chapter evaluates the impacts of the YOR19 campaign and examines the critical success factors of the campaign.

Homecoming Events in Emerging Destinations

There is always the urge for members of a diaspora community to go back to their ancestral homeland from which they migrated (Safran, 1991). This is also the case for even first- and second-generation immigrants who would also want to watch the sunset and then be buried in their native soil (Darieva, 2011) or to walk in the footsteps of their ancestors. According to Kasinitz et al. (2008, p. 258), as international tourism becomes more convenient and affordable, diasporic communities have more opportunities to visit their countries of origin, and what used to be a 'once-in-a-lifetime trip is now often an annual event'. Many governments have recognised the importance of such visits and are increasingly recognising their diasporas and welcoming them back home, usually through organised events (Sim & Leith, 2013). Emerging economies have particularly been at the forefront of organising homecoming events to attract their diaspora, as they recognise the importance of their diaspora in helping address their socio-economic challenges. China was one of the first countries to welcome its diaspora through tourism events. Since the 1980s, the Overseas Chinese Affairs Office has organised a number of tourism events to attract its diaspora from around the world and to encourage them to return to their homeland (Li & Mckercher, 2016). Also, regional governments such as the Guangdong government regularly organised tours to attract overseas Chinese to visit their villages to enhance local business partnerships and financial investments (Lew & Wong, 2003). Similar events have been organised by the district government of Lianhu of Xi'an in collaboration with the Cultural Centre of the Muslim Quarter since 2014 (Zhu, 2020). The Jamaican government also organises biannual conferences for the diaspora through its Ministry of Foreign Affairs and Foreign Trade. In 2011, for instance, a diaspora convention was organised under the theme 'One Nation: Jamaica and its Diaspora in Partnership' (Mortley, 2011). Apart from Ghana, a number of African countries including Senegal and Nigeria also stage such homecoming events to attract their diaspora. Senegal is on record one of the first African countries to have organised such an event when it hosted a number of high-profile African Americans during the African American Summit in 1995. Also, the Goree Island in Senegal regularly holds homecoming ceremonies for African Americans, which helps to market popular diaspora attractions such as *Juffure* and the 'house of slaves'. In Nigeria, the Island of Badagry, under the auspices of the Badagry local government, hosts a Black heritage festival to attract the African Diaspora. This event has been patronised by some high-profile African Americans including heavyweight boxing champion Evander Holyfield and the former Jackson Five member Marlon Jackson.

History of Ghana's Engagement with the African Diaspora

Ghana has always considered the African Diaspora as its niche market and has devoted significant attention to that market. Since gaining independence in 1957, successive governments have initiated programmes and activities geared towards attracting the African Diaspora to visit, stay, and invest in Ghana. President Nkrumah, after independence, positioned Ghana as the 'black star of Africa' and the territorial base of the pan-African movement. These efforts led to prominent African Americans such as Maya Angelou, Malcolm X, Richard Wright, Sylvia Boone, Julian Mayfield, and W.E.B. Du Bois visiting and staying in Ghana. The National Tourism Development Plan for Ghana (1996–2010) identified African Americans as an important niche market for roots tourism (GMOT, UNDP, & WTO, 1996). Pursuant to this, from the fourth republic (i.e., 1993 to date), successive governments have initiated a number of steps including staging events, granting citizenship rights and land grants to encourage diasporas to visit and stay in Ghana.

The Pan-African Historical Theatre Festival (PANAFEST) was introduced and celebrated biannually from 1991. Also, Emancipation Day and Juneteenth, which were already being celebrated in the African Diaspora, were adopted and celebrated in Ghana annually. President Rawlings, the originator of these events, played an instrumental role in persuading African Americans to invest in the country. The Ministry of Tourism was renamed the Ministry of Tourism and Diaspora Relations in 1993 to lend further credence to the government's commitment to attracting the African Diaspora. The ministry, in collaboration with other state agencies, initiated a number of programmes, including fora, conferences, festivals, and other events, which were geared towards encouraging the Diaspora's return to their 'homeland'. There was also a symbolic return of the mortal remains of two slave ancestors, Crystal from Jamaica and Samuel Carson from the United States to Ghana through the 'door of no return' of the Cape Coast Castle, and they were subsequently re-interred at Assin Manso. Ghana hosted events like the African African-American (AAA) summit in 1999.

Between 2003 and 2009, the government initiated several steps to position Ghana as a preferred destination for the African Diaspora. These included the passage of a dual citizenship bill in 2002 and the Representation of the People Amendment Bill (ROPAB) (Act 699) in 2000 and the 'Right of Abode' law of 2000, which allows a person of African descent to apply and be granted the right to stay in Ghana indefinitely. These initiatives aimed at encouraging and facilitating the involvement of the diaspora in the development of the country.

The government then initiated the Joseph Project in 2007 to encourage the African Diaspora to return to the homeland on a pilgrimage and to reconnect with their ancestors. President J.A. Kufuor also launched Africa's Golden Gateway Project in Washington in October 2005, which aimed at stepping up Ghana's efforts at attracting African American investors and tourists. The visit by the first African American President, Barack Obama, to Ghana in 2009 gave further credence to Ghana as a destination for the African Diaspora. However, these programmes and activities have achieved varying degrees of success. In 2016, 37 African Diasporas were granted Ghanaian citizenship. In 2019, during the YOR19, President Nana Akuffo Addo granted Ghanaian citizenship to another 126 African Diasporas.

The Year of Return 2019 (YOR19) Campaign

President Akuffo Addo launched the YOR19 campaign in Washington, D.C., in September 2018. At the launch, he indicated that:

> We know of the extraordinary achievements and contributions (Africans in the diaspora) made to the lives of the Americans, and it is important that this symbolic year—400 years later—we commemorate their existence and their sacrifices. (Egbejule, 16 September 2019)

The year-long homecoming event sought to position Ghana as a key travel destination for African Americans and the African Diaspora by boosting tourism while reconnecting African Diaspora to their roots. The YOR19 commemorated 400 years since the first enslaved Africans arrived in Jamestown, Virginia, in the United States. The first slave ship arrived in Point Comfort, Virginia, in August 1619. This arrival heralded the birth of slavery in America.

The year-long campaign was largely events driven, including conferences, investment fora, festivals, musical concerts, carnivals, sporting events, food and trade bazaars, fashion shows, visits to heritage sites, and other reverential events. The YOR19 was not limited to only cultural and leisure events, but business events geared towards encouraging investment by the diaspora in Ghana. However, most of the events were cultural events including Afrochella, Afro Nation, Afrochic Diaspora Festival, December in Ghana, PANAFEST, Emancipation Day, the Black Gala, Back to Africa Festival, *Chale Wote*, and *Potomanto* festival. In all, about 130 events were staged throughout the year.

Another significant feature of the YOR19 was the number of high-profile African American celebrities who graced the events. These included actors like Danny Glover, Idriss Elba, Michael Jai White, Boris Khodjoe and Samuel Jackson; musicians like Akon, Ludacris, Clifford Joseph Harris Jr. (T.I.), and Rick Ross; TV Host Steve Harvey, Senator Ilhan Omar, and wrestler Kofi Kingston. Over 40 African diaspora celebrities too attended the 'Full Circle Festival', which heralded the YOR19 events. In addition, prominent groups like the Congressional Black Caucus, which is made up of most African American members of the US Congress and *the National Association for the Advancement of Colored People* (NAACP), a civil rights organisation in the United States, sent delegations to the events.

Evidence provided by government sources suggests that the YOR19 was a resounding success. The Minister of Tourism, Arts and Culture, Mrs Barbara Oteng-Gyasi, indicated that the YOR19 campaign had generated a total of US$1.9 billion through economic activities and resulted in an increase in tourist arrivals of over 200,000 (*Daily Graphic* online, 2019). Additionally, tourism and hospitality businesses such as hotels, tour operators, restaurants, and handicraft shops have received a massive boost. The Ghana Tourism Authority (GTA) conducted a survey of attendees at the 'December in Ghana' event, and 35% of attendees indicated that they were satisfied with the event, while 54% indicated that they were very satisfied. Also, 97% had intentions of revisiting the country (GTA, 2020).

Impacts of the Year of Return Campaign

Improvement in Diaspora Relations

While the commitment of successive governments to strengthening the relationship between Ghana and the Africa Diaspora is not in doubt, the YOR19 campaign moved the relationship a notch higher. The YOR19 campaign brought the African Diaspora closer to Ghana and stimulated renewed interest by the diaspora in visiting and staying in Ghana. The granting of Ghanaian citizenship to 126 African Diasporas further strengthened the relationship as there was a recognition among the diaspora that Ghana was willing to accept them as one of their own. As a result, a record number of African Diaspora applied for citizenship during the YOR19 through the Ministry of Interior. By these acts, Ghana provided solace for African Diaspora who wanted to escape systemic racism in the diaspora and to be reunited with their kith and kin. That is the case of Dr Obadele Kambon, an African American lecturer who relocated to Ghana in 2008 but received Ghanaian citizenship during the YOR. According to him, since he resettled in Ghana, he no longer feels he is a victim of racial profiling or racial abuse (Nunoo, 24 August 2019). The euphoria and awareness that the YOR19 generated have rekindled a new sense of belongingness to Ghana as an ancestral homeland among African Diasporas. As a result, some African Diasporas have resettled in Ghana, while some prominent African Americans such as Stevie Wonder, Meek Mill, and Dave Chappelle have declared their intention to move to Ghana.

Increase in International Arrivals

A good measure of the success of an event is the number of visitors to the event. Data compiled by the GTA show that 1.13 million visitors travelled to Ghana in 2019 representing a growth of 18% over the previous year (Table 10.1).

Much of this growth is attributable to the YOR19 because that was the main event throughout the year. Specifically, tourist arrivals increased from 956,375 in 2018 to 1,130,307 in 2019. Tourist arrivals had dropped by 12,781 from 2017 to 2018. However, the YOR19 campaign helped to revive the sagging tourist arrivals in 2019, increasing by 18%. Also, arrivals from the United States, Ghana's niche market, increased by 10.4% from 2018 to 2019 (GTA, 2020). The number of visitors at the major tourist attraction sites also increased in 2019 as a reflection of the increase in tourist arrivals, as shown in Table 10.2. Arrivals at the Cape Coast Castle, which, according to Mensah (2015), is the main focus of roots tourism among African Diasporas, increased from 74,9987 in 2018 to 88,124 in 2019, a growth rate of nearly 15%.

The increase in tourist arrivals was also evident in the number of visa applications. According to the Ministry of Foreign Affairs, 800,000 visas had been issued as of December 2019. In anticipation of an increase in visa application in December during the climax of the YOR19, the government had to institute a visa-on-arrival programme. To encourage more people to visit the country, visa-on-arrival fees were reduced from US$150 to US$75. Also, the president announced a reciprocal visa waiver agreement with Jamaica (GTA, 2019).

Table 10.1 Tourist arrivals, receipts, and average expenditure (2016–2019)

Year	Arrivals	Receipts (M$)	Average expenditure[a]
2016	932,579	1651.67	1805.10
2017	969,156	1804.56	1862.00
2018	956,375	2589.85	2708.00
2019	1,130,307	3312.93	2931.00

Source: Ghana Tourism Authority (2020).
[a]Average expenditure is for the entire duration of the trip.

Table 10.2 Total annual arrivals at selected attraction sites (2016–2019)

	Year			
Site	2016	2017	2018	2019
Cape Coast Castle	74,440	61,228	74,987	88,124
Elmina Castle	44,574	47,306	65,208	69,544
Kwame Nkrumah Memorial	57,224	56,515	82,262	98,678
Manhyia Palace Museum	49,835	45,053	53,002	60,423

Source: Ghana Tourism Authority (2020).

Increase in Tourism Receipts and Stimulation of Economic and Social Activities

The increase in tourist arrivals had a concomitant effect on tourism receipts. As can be seen in Table 10.1, tourism receipts rose from US$2589.85 in 2018 to US$3312.93 in 2019. During the same period, the average tourist expenditure rose from US$2708 to US$2931. This was largely due to the YOR19 events. An increase in the average length of stay in the country from eight days to ten days also explains the increase in average tourist expenditure, as all other things being equal, the longer the stay, the greater the expenditure. The increase in expenditure had a ripple effect on other economic activities, especially those connected to the tourism industry such as airlines, handicrafts, tour operations, food and beverage services, transportation, and accommodation. Handicraft sellers at the Arts Centre confirmed an increase in sales, and in the last week of December, which also coincided with the Christmas festivities, hotels in Accra were recording 100% occupancy. The rise in economic activities also provided business and employment opportunities for some Ghanaians. For instance, one of the major events that climaxed the YOR19, the Afro Nation festival, gave temporary employment to 150 labourers daily throughout the four days of the event. In addition, the event provided business opportunities for some local real estate companies that provided accommodation for executives and artists billed for the festival (Brown, 18 January 2020).

There were also some socio-economic benefits resulting from the corporate social responsibility projects of the YOR19 secretariat and other corporate bodies. For instance, the GTA built a tourism information centre and memorial garden as symbols of the YOR19 at Anomabu. Also, some of the returnees undertook projects in some local communities, including the construction of a school at Akosombo. There were other socio-economic benefits in the form of investments, acquisition of properties and remittances from the diaspora.

Increase in International Media Coverage and Awareness of the Country

Apart from the above-mentioned tangible benefits of the YOR19 campaign, it also enhanced the image of Ghana by placing the media spotlight on the country. According to the GTA, YOR19 was covered domestically and internationally over 1000 times by television, print, radio, and online media sources. The Chief Executive Officer of the GTA, Akwesi Agyemang, said:

> It has been like a miracle if you look at even the media aspect of this; having Aljazeera, BBC, CNN, Ebony Magazine, Essence and all of them here, we even had Buzzfeed coming in to do another documentary, we couldn't have paid for the media. (Nyavor, 31 December 2019)

Some media personalities like Steve Harvey and Conan O'Brien visited the country and filmed special documentaries. Also, Roland Martin interviewed President Akuffo Addo on the YOR19. This media coverage helped in projecting Ghana as a preferred destination for the African Diaspora.

Critical Success Factors of the YOR19

There are some critical success factors (CSFs) that accounted for the successes achieved during the YOR19, and these have far-reaching implications for the marketing of emerging destinations. In order for such destinations to compete effectively, it is imperative that destination marketers identify these CSFs and ensure they are incorporated into the overall destination marketing plan (Baker & Cameron, 2008). The CSFs of the YOR19 include collaboration, celebrity endorsement, effective use of funds, and the easing of entry restrictions.

Collaboration

Collaboration is defined as an interactive process through which 'autonomous stakeholders of a problem domain use shared rules, norms, and structures to decide on issues related to that problem domain' (Wood & Gray, 1991, p. 146). The YOR19 campaign was effectively executed with the active collaboration of key stakeholders. The sheer scale and volume of events required active stakeholder engagement to ensure success. There was a collaboration among stakeholders including the Ghana Tourism Authority, Diaspora Affairs Office of the President, Ghana Tourism Federation, the Adinkra Group (USA), PANAFEST Foundation, W.E.B. Du Bois Centre, African American Association of Ghana, and, at the policy level, the Ministry of Tourism, Arts and Culture as well as the Ministry of Foreign Affairs. The steering committee was, thus, composed of representatives from various stakeholders. Collaboration also ensured that the events were largely organised by private sector organisations. There was a public call for proposals from organisers, leading to the granting of approval for some events to be held. More than 250 event proposals were submitted, out of which 130 were approved. In most cases, once an event was approved, the organisers had to find their own sponsors. However, the YOR19 Secretariat assisted with the promotion of the events. This form of collaboration also ensured that the individual stakeholders worked collectively to meet the needs of visitors and to the benefit of the entire destination. This is particularly imperative due to the composite nature of the tourist product.

Celebrity Endorsement

An important strategy that was employed in the promotion of the YOR19 campaign was celebrity endorsement. There is the tendency of using film actors, sports stars, fashion models, and other celebrities to promote products and destinations because they are able to penetrate the commercial clutter of advertising and gain consumer attention (Magini et al., 2008; Morgan & Pritchard, 2001). The events were spearheaded by celebrities mainly from the United States (i.e., the niche market), and some local celebrities. Hollywood actor Boris Kodjoe and marketing icon Bozoma Saint John, both of Ghanaian parentage, set the tone when they organised the Full Circle Festival to herald the year-long event. They were able to draw a number of high-profile celebrities including Idris Elba, Naomi Campbell, Boris Kodjoe, Michael Jai White, Anthony Anderson, and Rosario Dawson to the event. Boris Kodjoe in an interview on CNN said, 'Every person of colour needs to get on this pilgrimage. They need to experience this journey and get in touch with their emotional heritage, walk through the dungeons and see the 'door of no return'' (Busari & Siaw, 7 February 2019). Most of the celebrities who have a huge following on social media, including Cardi B, Ludacris, and Steve Harvey, shared their experiences on social media, and this ignited a new wave of interest in visiting Ghana.

For instance, when President Trump tweeted that some congressional women should 'go back' to 'the totally broken and crime-infested places from which they came', Ilhan Omar, a US congresswoman who is originally a Somalian, responded with a tweet from the Cape Coast Castle in August when she visited as part of the Congressional Black Caucus delegation: 'So grateful for the honour to return to Mother Africa' (Paquette, 27 December 2019). This tweet created a frenzy among many African Diasporas who wanted to visit the country.

Effective Use of Funds

Since destination marketing campaigns usually involve a significant investment in promotional activities, the success of such campaigns also hinges on a good return on investment. This is particularly imperative because, in many destinations, public sector funding of the tourism sector has come under scrutiny (Fyall et al., 2007). The socio-economic benefits derived from such campaigns must justify the expenditure.

However, destinations need to know whether visitor numbers and spending can be linked directly to marketing campaign activity (Pratt et al., 2010). According to the Chief Executive Officer of the Authority, Akwesi Agyemang, a total amount of GH¢6 million (almost US$1 million) was spent on the YOR19 campaign both locally and internationally (Myjoyonline.com, 2019). This is a very modest amount by every standard. For instance, the Homecoming Scotland campaign in 2009 cost £5.5 million in marketing expenditure and £3 million on programme of activities, culminating in a total expenditure of £8.5 million (Morrison & Hay, 2010). Also, Rwanda spent about £30 million for its 'Visit Rwanda' sleeve sponsorship with English Premier League side, Arsenal, in 2018, and between US$9 million and $11 million on another sponsorship deal with the women's team of Paris Saint-Germain (PSG) (Reuters, 2019). While the circumstances of these expenditures may not be the same as the YOR19, it is still remarkable, considering the impact of the YOR19. Unlike the HS09 campaign, the events were mainly organised by the private sector, while the YOR19 Secretariat focused more on promotional activities. There were a number of cost-cutting measures. The organisers were able to keep the marketing expenditure down because they did not spend much on documentaries and advertisements in the international media but employed the influencer and social media marketing strategy. Additionally, the celebrities who endorsed the country were not paid, and they visited the country at their own expense. The GTA and the YOR19 Secretariat employed the appeal of celebrities, the popularity of social media, and a public relations approach to lure African Diasporas to visit the country.

Easing of Entry Restrictions

Another factor that accounted for the success of the YOR19 was the easing of restrictions on entry into the country. The government eased entry visa restrictions in order to boost tourist arrivals. The government instituted the visa-on-arrival programme in December 2019 and constituted steering committees both locally and abroad to ensure the speedy issuance of visas to Ghana. Also, visa-on-arrival fees were reduced from

US$150 to US$75, to enable more people to visit the country. Moreover, the president also announced a reciprocal visa waiver agreement with Jamaica. Following the Right of Abode law in 2000, which grants people of African descent the right to apply and be granted the right to stay in Ghana indefinitely, the president granted Ghanaian nationality to some 126 diasporas during the YOR19.

Beyond the Return

In order to build on the success of the YOR19 and consolidate the benefits to the country and the African Diaspora, President Nana Akufo-Addo, on Friday 27 December 2019, launched the 'Beyond the Return, The Diaspora Dividend' campaign as a sequel to the YOR19 campaign. During the launch, he called on Ghanaians to engage Africans in the diaspora and persons of African descent more positively in areas such as trade and investment cooperation, skills and knowledge, and development. He further explained that the 'Beyond the Return' (BTR) was to ensure that Ghana derived maximum benefits from its relations with the diaspora through mutually beneficial cooperation and partnership for shared growth and development. BTR, which will span a decade (2020–2030), is founded on seven pillars:

- Experiencing the culture, warmth, and rhythm of the country (Experience Ghana)
- Providing diverse investment opportunities (Invest in Ghana)
- Easing the travel experience (Diaspora Pathways to Ghana)
- Promoting cultural festivals and events (Celebrate Ghana)
- Branding Ghana as the hub of African renaissance (Brand Ghana)
- Support for communities (Give Back to Ghana)
- Promotion of roots tourism (Promote Pan-African Heritage and Innovation)

However, BTR has been hampered by the COVID-19 pandemic. Many planned events had to be cancelled in 2020 due to the imposition of restrictions including the closure of the Kotoka International Airport. As a result, the government had to revise its plans. However, with the gradual easing of restrictions in 2021, it is expected that the BTR programme will bounce back.

Conclusion

Ghana has continued to engage with its diaspora since independence. Successive governments have initiated a number of programmes and activities to position the country as the preferred destination for the African Diaspora, particularly the African American niche market. The YOR19 was, however, the most significant campaign by the government targeted at the diaspora. This was a series of events in 2019 that was heavily patronised by the diaspora. The YOR19 provides a classic case of how an emerging economy can employ homecoming events to attract its diaspora. The government is capitalising on the success of the YOR19 to develop a decade-long series of activities and events known as 'Beyond the Return'. However, for the country to reap the full benefits of the BTR, there is the need to undertake a thorough evaluation of the YOR19 event, as this will provide valuable lessons on how to deal with the African Diaspora and how best to serve their needs. Following the success of the YOR19 campaign, other African countries, notably Nigeria and Benin, are planning similar events. There are already signs of heightened competition for a share of the African Diaspora market. Ghana must, therefore, preserve its competitive position by developing well-thought-through products and promotional strategies and activities targeted at the African Diaspora.

Acknowledgements The author wishes to acknowledge Mr Jones Nelson, Acting Director of Corporate Affairs, and Mr Spencer Doku, Manager of the Research Monitoring and Evaluation Department of the Ghana Tourism Authority for supplying information and data for this chapter.

References

Ankomah, P., Larson, T., Roberson, V., & Rotich, J. (2012). A creative approach to development: The case for active engagement of African diaspora in Ghana. *Journal of Black Studies, 43*(4), 385–404.

Asiedu, A. (2005). Some benefits of migrants' return visits to Ghana. *Population, Space and Place, 11*(1), 1–11.

Baker, M. J., & Cameron, E. (2008). Critical success factors in destination marketing. *Tourism and Hospitality Research, 8*(2), 79–97.

Basu, P. (2005). Macpherson country: Genealogical identities, spatial histories and the Scottish diasporic clanscape. *Cultural Geographies, 12*(2), 123–150.

Brown, O. (2020, January 18). Afro nation Ghana festival: Tourism and economic benefits. *Citinewsroom*. https://citinewsroom.com/2020/01/afro-nation-ghana-festival-tourism-and-economic-benefits-article/

Busari, S., & Siaw, L. (2019, February 7). Ghana is being heralded as the next big tourist destination. Here's why. *CNN Inside Africa*. https://edition.cnn.com/travel/article/ghana-year-of-return-tourism-intl/index.html

Daily Graphic online. (2019, December 16). *Year of return generates $1.9 billion*. https://www.graphic.com.gh/news/general-news/year-of-return-generates-1-9-billion.html

Darieva, T. (2011). Rethinking homecoming: Diasporic cosmopolitanism in post-Soviet Armenia. *Ethnic and Racial Studies, 34*(3), 490–508.

Egbejule, E. (2019, September 16). Diaspora tourists heed the call of Ghana's 'Year of Return'. *The Africa Report*. https://www.theafricareport.com/17265/diaspora-tourists-heed-the-call-of-ghanas-year-of-return/

Fyall, A., Fletcher, J., & Spyriadis, T. (2007). *Diversity, devolution and disorder. The management of destinations in England*, 2007. Advances in Tourism Marketing Conference, Valencia, Spain, September 10–12.

Fyall, A., Garrod, B., & Tosun, C. (2006). Destination Marketing: A framework for future research. In M. Kozak & L. Andreu (Eds.), *Advances in tourism research: Progress in tourism marketing* (pp. 75–86). Elsevier.

Getz, D. (1997). *Event Management and event tourism*. Cognizant Communications Corporation.

Ghana Tourism Authority. (2019). *Year of return in perspective part 1: A presentation on the impact of the year of return*. GTA.

Ghana Tourism Authority. (2020). *2019 tourism report*. GTA.

GMOT, UNDP, & WTO. (1996). *National tourism development plan for Ghana (1996–2010). Integrated tourism development programme (GHA/92/013).*

Ghana Ministry of Tourism, United Nations Development Program, and World Tourism Organization.
Kasinitz, P., Mollenkopf, J. H., Waters, M. C., & Holdaway, J. (2008). *Inheriting the city.* .
Kleist, N. (2013). Flexible politics of belonging: Diaspora mobilisation in Ghana. *African Studies, 72*(2), 285–306.
Lew, A. A., & Wong, A. (2003). News from the motherland: A content analysis of existential tourism magazines in Southern China. *Tourism Culture & Communication, 4*(2), 83–94.
Li, T. E., & McKercher, B. (2016). Developing a typology of diaspora tourists: Return travel by Chinese immigrants in North America. *Tourism Management, 56*, 106–113.
MacGonagle, E. (2006). From dungeons to dance parties: Contested histories of Ghana's Slave Forts. *Journal of contemporary African Studies, 24*(2), 249–260.
Magini, V. P., Honeycutt, E. D., & Cross, A. M. (2008). Understanding the use of celebrity endorsers for hospitality firms. *Journal of Vacation Marketing, 14*(1), 57–69.
Mensah, I. (2015). The roots tourism experience of diaspora Africans: A focus on the Cape Coast and Elmina Castles. *Journal of Heritage Tourism, 10*(3), 213–232.
Mensah, I. (2021). Roots tourism. In D. Buhalis (Ed.), *Encyclopedia of tourism management and marketing.* Edward Elgar. https://doi.org/10.4337/9781800377486.roots.tourism
Mensah, I., & Amissah, E. (2009). The prospects and challenges of positioning Ghana as a preferred African American tourist destination. In A. Fyall, M. Kozak, L. Andreu, & S. S. Lebe (Eds.), *Marketing innovations for sustainable destinations* (pp. 203–215). Goodfellow.
Morgan, N., & Pritchard, A. (2001). *Advertising in tourism and leisure.* Butterworth-Heinemann.
Morgan, N., Pritchard, A., & Pride, R. (2003). Marketing to the Welsh diaspora: The appeal to hiraeth and homecoming. *Journal of Vacation Marketing, 9*(1), 69–80.
Morrison, A., & Hay, B. (2010). A review of the constraints, limitations and success of Homecoming Scotland 2009. *Fraser of Allander Economic Commentary, 34*(1), 44–54.
Mortley, N. K. (2011). Strategic opportunities from diaspora tourism: The Jamaican perspective. *Canadian Foreign Policy Journal, 17*(2), 171–185.

Myjoyonline.com. (2019, December 30). *GH¢6m spent on 'Year of Return' campaign—Tourism Authority.* https://www.myjoyonline.com/news/2019/December-30th/gh6m-spent-on-year-of-return-campaign-tourism-authority.php

Nunoo, F. (2019, August 24). The African American who moved to Ghana 'to escape US racism'. *BBC Pidgin.* https://www.bbc.com/news/world-africa-49394354

Nyavor, G. (2019, December 31). 'It's more than I expected'—Akufo-Addo impressed by year of return results. *Myjoyonline.* https://www.myjoyonline.com/news/2019/december-31st/its-more-than-i-expected-akufo-addo-impressed-by-year-of-return-results.php

Paquette, D. (2019, December 27). How Trump's tweets and Cardi B are making more Americans move to Ghana. *Washington Post.* https://www.washingtonpost.com/world/africa/how-trumps-tweets%2D%2Dand-cardi-b%2D%2Dare-making-more-americans-move-to-ghana/2019/12/26/f3635476-1dfc-11ea-977a-15a6710ed6da_story.html?outputType=amp

Poirier, R. A. (2000). A Tourism in the African economic milieu: A future of mixed blessing. In P. U. C. Dieke (Ed.), *The political economy of tourism development in Africa* (pp. 29–36). Cognizant Communication Corporation.

Pratt, S., McCabe, S., Cortes-Jimenez, I., & Blake, A. (2010). Measuring the effectiveness of destination marketing campaigns: Comparative analysis of conversion studies. *Journal of Travel Research, 49*(2), 179–190.

Reuters. (2019, December 12). Rwanda signs deal with Paris St Germain to promote tourism. *Euronews.* https://www.euronews.com/2019/12/04/rwanda-signs-deal-with-paris-st-germain-to-promote-tourism

Ritchie, J. R. B., & Crouch, G. I. (2003). *The competitive destination: A sustainable tourism perspective.* CABI.

Safran, W. (1991). Diasporas in modern societies: Myth of homeland and return. *Diaspora, 1*(1), 83–99.

Sim, D., & Leith, M. (2013). Diaspora tourists and the Scottish Homecoming 2009. *Journal of Heritage Tourism, 8*(4), 259–274.

Wood, D. J., & Gray, B. (1991). Toward a comprehensive theory of collaboration. *The Journal of Applied Behavioral Science, 27*(2), 139–162.

Zhu, Y. (2020). Memory, homecoming and the politics of diaspora tourism in China. *Tourism Geographies*, 1–18.

11

Destination Branding Slogans and their Impact on Tourist Arrivals: The Case of the Philippines

Reil G. Cruz and Joy Sheelah B. Era

Introduction

In 2019, travel and tourism generated US$9 trillion, representing more than 10% of the world's gross domestic product (World Travel and Tourism Council, 2020). It also generated more than 334 million jobs in the same year, equivalent to one in every ten jobs worldwide. The industry's share of the GDP of the Philippines is even greater at almost US$91 billion, or more than 25%. At the same time, travel and tourism sustained more than 10 million jobs or 24% of the country's total employment. No

R. G. Cruz (✉)
University of the Philippines Asian Institute of Tourism,
Quezon City, Philippines
e-mail: rgcruz@up.edu.ph

J. S. B. Era
Institute of Tourism and Hotel Management, Far Eastern University,
Manila, Philippines
e-mail: jsera@feu.edu.ph

I. Mensah et al. (eds.), *Marketing Tourist Destinations in Emerging Economies*, Palgrave Studies of Marketing in Emerging Economies,
https://doi.org/10.1007/978-3-030-83711-2_11

wonder then that several countries have made tourism a pillar of their economies. However, with a total of 195 countries, tourism is a hyper-competitive industry. Countries must 'cut through the clutter of marketing communication messages' to reach the minds of the target tourist segments (Pike, 2004, p. 4). Market positioning requires a 'succinct message' that focuses on one or a few brand associations (Pike, 2004, p. 4). This message is encapsulated in the brand name as a logo and a slogan (Pike, 2004). Countries have used slogans as part of their overall destination branding strategy, underscoring their importance as a competitive instrument. However, the topic of destination brand slogan remains a murky area in tourism literature (Pike, 2004). Moreover, studies that investigate the impact of destination brand slogans on a country's tourist arrivals remain a huge gap in destination brand scholarship, with practically all of them being conceptual in nature. This study attempts to shed light on this relationship by asking the question, 'Do slogans improve a country's tourism performance?' To answer this question, we review the destination brand slogans of the Philippines from 1973 to the present and correlate them to the country's tourist arrivals during the same period.

The Evolving Concept of Destination Brand

The concept of destination brand originated from concepts borne in the marketing discipline. The term, *brand*, refers to the collection of perceptions that a person has of a product (or destination) based on the product's (or destination's) observable characteristics, which differentiate it from all other products (or destinations), thereby endowing it with a competitive advantage over other similar products (or destinations) (World Tourism Organization & European Travel Commission [WTO & ETC], 2009). The WTO and ETC (2009) use brand and competitive identity as synonyms. There are several other concepts that are built around the brand, as defined by the WTO and ETC.

Brand architecture is the way in which different brands within a destination are structured, how they relate to each other, and how they are differentiated from each other. The different brands should demonstrate a strong 'family' resemblance but still are individually distinctive. Brand attributes are the 'rational' attributes of a destination—the things that visitors like to do and see. They are objectively identifiable and not endowed with any emotional connotation, and contribute towards the destination's distinctiveness. Brand equity is defined as 'the set of brand assets and liabilities linked to a brand, its name, and symbol that add or subtract from the value of a product or service to a firm or to its customers' (Aaker, as cited in Tran et al., 2019, p. 706). Brand equity has four dimensions: (a) brand awareness, (b) perceived quality, (c) brand association, and (d) brand loyalty (Aaker, as cited in Tran et al., 2019). Brand awareness is the quickness to which a customer can retrieve the elements of a brand stored in their memory over time (Mowen & Minor, as cited in Chigora & Zvavahera, 2015). Brand awareness can be broken down into top-of-the-mind awareness and added awareness. The former is the automatic association made by a customer between the brand and a particular product category, and the latter is the tendency of a buyer to pick a known brand when given a variety of brands (Farris et al., as cited in Chigora & Zvavahera, 2015). Perceived quality is defined as the elements that make people visit a particular tourism destination; it is a subjective judgement of overall quality (Ekinci & Hosany, as cited in Kim & Lee, 2018; Morgan et al., as cited in Kim & Lee, 2018). Brand association is defined as everything related to the memory of consumers about a brand and what that brand means to consumers (Keller, as cited in Listiana, 2015). Brand loyalty refers to a customer's tendency to repurchase a brand, indicated by repeated purchases over time (Bae et al., 2020). Brand essence refers to the functional and emotional benefits that tourists experience when they visit a destination (Ezeuduji & Nkosi, as cited in Mlambo & Ezeuduji, 2020, p. 1196). Brand identity is defined as 'a bundle of mental and functional associations about the brand providing familiarity and differentiation that are not replicable by other competitors; it poses the question as to how a tourist destination wants to be perceived by travellers (Dyankov et al., 2018, p. 21). Logo, a visual design mark used consistently on all destination marketing communications to symbolize the destination, is one of the tools that help create this identity (WTO & ETC, 2009). It gains recognition as representing a destination when used over time. Together with the logo, another essential component of destination brand identity is the slogan, defined as 'a short phrase used to convey descriptive and persuasive information about a particular [destination]' (Galí et al., 2016, p. 1 cited in Pan, 2019). A slogan impacts a brand personality, which influences a visitors' intention to recommend a destination (Ekinci & Hosany, as cited in Pan, 2019). Brand personality is the summation of the destination's character, the way the destination is perceived and reflects people's attitudes towards it (WTO & ETC, 2009). According to Qu et al. (2021), an effective tourism slogan should contain a short text, simple words, symmetric sentence structure, neat rhyme, and coordinated tone. The slogan must also have rich rhetoric and interactive information, including the destination name, have one or few focused propositions and offer unique tourist benefits (Qu et al., 2021).

An important concept in branding is brand positioning. Positioning is defined as 'an act of designing a company's offering and image so that they occupy a meaningful and distinct competitive position in the target market's minds' (Kotler, as cited in Saqib, 2020, p. 9). A part of the destination brand is the slogan (WTO & ETC, 2009). Pike (2004) proposed six questions or criteria for developing destination brand positions:

1. What is the slogan's value proposition?
2. Who will find the proposition meaningful?
3. How does the slogan differentiate the destination from that in the competitive set?
4. How is the message likely to be memorable?
5. Has the proposition been designed for the long term?
6. Is the proposition promise deliverable by the destination community?

Value proposition is the reason why a tourist would visit a destination (Kotler, 1999). The positioning could be based on an attribute, benefit, use/application, user, computer, category, or quality/price (Kotler, 1999).

Destination refers to a country, region or city that people wish to visit (WTO & ETC, 2009). Thus, a destination brand refers to the collection of perceptions that a person has of a destination based on the destination's observable characteristics, which differentiate it from all other destinations. It is also referred to as the destination's DNA—its core defining characteristics (WTO & ETC, 2009).

A destination brand may sometimes be confused with place brand, country brand, or national brand. However, the latter three are bigger concepts that cover tourism and several other sectors, such as manufacturing, inward investment and exports, sports, education, and entertainment (WTO & ETC, 2009). The place brand should convey a 'sense of place', the overall vibe of a place that is connected to its physical appearance, and cultural heritage consisting of its built and natural environment, the events it mounts, the products it produces, and the behaviour of its citizens. There should be a strong connection between the destination brand and the overall place brand (WTO & ETC, 2009). Tourism can impact the overall national brand image because holiday experiences can change preconceptions about a country. Tourists can become viral agents for disseminating images through word of mouth (Anholt, as cited in WTO & ETC, 2009). It follows that any government must provide satisfying tourism experiences and manage its communication activities to enhance these acquired images of the country.

Destination branding is defined as 'a multidimensional assortment of functional, emotional, relational, and strategic elements' (Aaker, as cited in Ruiz-Real et al., 2020). Destination branding is also defined as:

> a name, symbol, logo, wordmark or other graphic that both identifies and differentiates the destination: furthermore, it conveys the promise of a memorable travel experience that is uniquely associated with the destination: it also serves to consolidate and reinforce the recollection of pleasurable memories of the destination experience. (Ritchie & Ritchie, p. 18 in Castañeda-Garcia et al., 2020)

Ooi (as cited in Ruiz-Real et al., 2020) defines destination branding as the process of attracting the support and cooperation of various stakeholders so that the brand is accepted, communicated, and manifested through official and unofficial means.

Applying marketing techniques to destinations entails several difficulties due to the peculiar nature of destinations as brands (Ruiz-Real et al., 2020). First, destinations are more complex than consumer goods (Pike, 2004), since they represent numerous autonomous service enterprises, both public and private, where control over the brand experience is quite low (Hankinson, as cited in Ruiz-Real et al., 2020). In addition, destination branding entails providing experiences calibrated to different tourist markets (Gartner, as cited in Ruiz-Real et al., 2020). Destination branding involves hard and soft factors (Morgan et al., as cited in Ruiz-Real et al., 2020). Hard factors include infrastructure and the economy, while soft factors cover the environment, art, and culture.

Another related concept is destination brand architecture. It is described as 'the synergy that is achieved when sub-national brands are related to the national brand and are part of the same 'family'' (WTO & ETC, 2009, p. 60). Synergy results when the national brand values resonate in the sub-national brands (WTO & ETC, 2009). Like children, each sub-national brand could be distinct from the other but must resemble the national brand. Sub-national brands should use the national brands' websites and brochures to penetrate markets and widen distribution (WTO & ETC, 2009). Harish (2010) cites the case of brand architecture as a means to organize the brand portfolio that describes the brand roles and relationships between brands. In other words, brand architecture describes the congruence among the brands at different levels. The literature highlights its importance in effectively managing a brand from a firm, company, or organization's perspective. Aaker and Joachimsthaler (as cited in Harish, 2010) identify four principal types of brand architecture.

Harish (2010) identifies four types of destination brand architecture strategies:

1. Branded house: This is a brand used in all of its businesses, products, and services.
2. House of brands: This refers to the extensive range of stand-alone brands, often more than once in the same product category.
3. Sub-brands: It is a dual approach that uses both the company and product brand. If the company plays a dominant role, it is a sub-brand strategy.
4. Endorsed brands strategy: It is a dual approach in which the product plays a dominant role, and the company provides only a token endorsement.

In his work, Harish, 2010 cites the application of branded house strategy by New Zealand, house of brands strategy by Spain, sub-brands strategy by Western Australia, and endorsed brand strategy by Scandinavia. An illustration of the concept of family resemblance in destination brand architecture is provided in the *Handbook on Destination Branding* (WTO & ETC, 2009). Zululand is used as the regional brand. Each of the sub-regions uses a branding that is distinct but connected to the family brand. Examples are Durban—'Playground of the Zulu Kingdom', North Coast—'Jewel of the Zulu Kingdom' and South Coast—'Paradise of the Zulu Kingdom'. Each of the sub-regions offers different tourism experiences, urban vibe, beach, fun, and sport for Durban; lifestyle and pristine beaches for North Coast; and family beach holiday and family fun for South Coast.

Frameworks and Approaches

Destination brands can be examined through different lenses. As shown in Table 11.1, a large majority of the frameworks used in the study of destination branding have been formulated in the 1990s and early 2000s, and new approaches have been quite sparse. Many of the models originate from concepts in the marketing discipline. Examples of these are the brand equity model (Aaker, as cited in Almeyda-Ibañez & George, 2017), unique selling proposition concept (Richardson & Cohen, as cited in Almeyda-Ibañez & George, 2017), information economics perspective (Erdem & Swait, as cited in Almeyda-Ibañez & George, 2017), service branding model (Berry, as cited in Garcia et al., 2012), and customer-based brand equity (CBBE) (Keller, as cited in Almeyda-Ibañez & George, 2017). Some of the more novel approaches to destination branding analysis include the brand architecture approach (Harish, 2010), strategic place brand management (SPBM) model (Hanna & Rowley, 2011), message consistency model (Castañeda-Garcia et al., 2020), and stakeholder-based destination branding model (Garcia et al., 2012). Stakeholders could include local people, investors, employees, students, retired people, visitors, media and opinion leaders, tourism entrepreneurs, service industries, foreign governments, and exports purchasers

Table 11.1 Framework for destination branding analysis

Brand Equity Model (Aaker, 1991 in Almeyda-Ibañez & George, 2017)	Destinations may be examined from the four assets that comprise brand equity: brand awareness, perceived quality, brand associations and brand loyalty
Unique Selling Proposition (Richardson & Cohen, 1993 in Almeyda-Ibañez & George, 2017)	A hierarchical scale for analysing states' marketing slogans, which ranged from 'Level 0: No proposition' through 'Level 4b: Unique selling proposition'
Information Economics Perspective (Erdem & Swait, 1998 in Almeyda-Ibañez & George, 2017)	Focuses on the role of credibility as the main determinant of customer-based brand equity The content, clarity, and credibility of a brand may increase the consumers' perceived quality of the brand and reduce the information costs and perceived risks
Service Branding Model (Berry, 2000 in Garcia et al., 2012)	Examines tourist destinations from six main components: presented brand, external brand communications, customer experience with the organization, brand awareness, brand management, and brand equity
Over-all Brand Equity (Yoo et al., 2000 in Castañeda-Garcia et al., 2020)	Evaluates a destination from a multidimensional BE scale, which includes brand awareness, perceived quality, brand associations, and brand loyalty
Destination Branding Model (Cai, 2002)	Revolves around an axis formed by brand mix (slogans or logos). The other components are brand identity and brand image building
Destination Brand Management Strategies	Compares the advantages and disadvantages of (1) single-destination brand; (2) umbrella destination brand; and (3) multiple-destination brands
Customer-Based Brand Equity (CBBE) (Keller, 2002 in Almeyda-Ibañez & George, 2017)	A brand has positive customer-based brand equity when customers react more positively to the product and to its marketing tactics when the brand is identified, as compared to when it is not
Competitive Identity Hexagon (Anholt, 2002 in WTO & ETC, 2009)	Looks at the relationships among the six areas of tourism, products and service brands, government policies, investment, cultural exchange, cultural exports and sport, and people of the country

(*continued*)

Table 11.1 (continued)

Brand Equity Model (Aaker, 1991 in Almeyda-Ibañez & George, 2017)	Destinations may be examined from the four assets that comprise brand equity: brand awareness, perceived quality, brand associations and brand loyalty
Brand Pyramid (WTO & ETC, 2009)	A brand-building model for National Tourism Organizations to use in building a destination brand. It consists of the following sequential elements to build up the brand: rational attributes, emotional benefits, brand personality, positioning statement, and brand essence/brand values.
Brand Wheel (WTO & ETC, 2009)	A brand-building model is a useful technique for developing a destination brand. It depicts the following elements on a wheel-shaped diagram: values, personality, benefits, substantiators (facts about the brand that support the proposition), essence, and proposition
Conceptual Model of Destination Branding (Hsu & Cai, 2009)	This model is divided into brand knowledge, brand trust and brand loyalty
Strategic Place Brand Management (SPBM) Model (Hanna & Rowley, 2011)	Key components of the model are brand evaluation, stakeholder engagement, infrastructure, brand identity, brand architecture, brand articulation, marketing communications, brand experience, and word of mouth
Place Branding Framework (Tasci, 2011 in Adeyinka-Ojo & Nair, 2016)	The approach cites different levels of branding: country branding, nation branding, destination branding, or smaller-level city branding. It is based on the relationships among destination management and marketing, destination brand, and destination brand equity
Brand Architecture Approach (Harish, 2010)	Looks at the relationship and resemblance of alternative destination strategies ranging from a national or umbrella brand to sub-brands
Stakeholder Based-Destination Branding Model (Garcia et al., 2012)	Uses the Success Index of Triple-Diamonds (SITD) to evaluate a destination brand's success from the point of view of stakeholders, such as local people and business owners

(continued)

Table 11.1 (continued)

Brand Equity Model (Aaker, 1991 in Almeyda-Ibañez & George, 2017)	Destinations may be examined from the four assets that comprise brand equity: brand awareness, perceived quality, brand associations and brand loyalty
Consumer/Customer-Based Destination Brand Equity (Bianchi et al., 2014; Boo et al., 2009)	Looks at destinations from the perspectives of tourists along the following dimensions: brand image; brand quality; brand awareness; brand loyalty; and perceived value
Brand Conceptual Model (Kotler, 2016)	Consists of the six elements (brand purpose, brand positioning, brand differentiation, brand identity, brand trust, and brand beneficence)
Brand Activism Model (Kotler & Sarkar, 2017)	Looks at how destinations support any social legal, business, economic, political, and environmental issues
Technologies of Power Approach (Lund et al., 2018)	Destinations are examined through the concepts of storytelling, mobilities, performances, and performativities
Message Consistency Model (Castañeda-Garcia et al., 2020)	Analyses the effect of these two variables on CBDBE

(Gilmore, as cited in Garcia et al., 2012). The 'technologies of power' (Foucault, as cited in Lund et al., 2018) framework looks at the use of social media by people to exert influence on others, utilizing the concepts of storytelling, mobilities, performances, and performativities. This framework advises destination marketing organizations to recognize the importance of collaborative storytelling, co-creating stories with the users, and regard social media users as potential allies.

Kotler's branding conceptual model (2016) consists of the following steps:

1. Brand purpose
2. Brand positioning
3. Brand differentiation
4. Brand identity
5. Brand trust
6. Brand beneficence

According to Kotler (2016), brand purpose does not mean functional purpose but also a higher purpose. The higher purpose refers to the emotional and social benefits that come with choosing a particular brand. Thus, to provide an example, an airline's functional purpose is to fly people from origin to the destination, but its higher purpose is to connect loved ones and family or make passengers feel at home. Brand trust means a brand's reliability so that its customers believe that the brand will deliver what it claims. For example, a flight schedule is something airlines try to observe because that is what the passengers expect. The last concept and perhaps the most important one is brand beneficence, or how well a brand serves both its customers and the society at large. Companies may do this by adhering to ethical business practices and sponsoring their own corporate social responsibilities (CSRs). In recent years, the concept of CSR has evolved into the term 'brand activism'. Brand activism could come in many forms, such as social activism (equality—gender, LGBT, race), legal activism (tax, workplace, and employment laws), business activism (CEO pay, worker compensation, labour and union relations, governance), economic activism (minimum wage and tax policies), political activism (lobbying, voting rights, and policy), and environmental activism (conservation, environmental, land-use, pollution laws) (Kotler & Sarkar, 2017). Destinations may be involved in some form of environmental conservation or sustainability issues without even knowing that they are actually engaging in some form of brand activism.

Success Factors

Building a successful destination brand involves several critical factors. From the literature, these factors include having a good understanding of the destination's core market segments, qualitative consumer research, long-term commitment for the brand to gain traction in the market (WTO & ETC, 2009), a unique selling proposition (Richardson & Cohen, as cited in Almeyda-Ibañez & George, 2017), strong brand equity, brand awareness and familiarity (Keller, as cited in

Almeyda-Ibañez & George, 2017), regular monitoring (Garcia et al., 2012), message consistency (Castañeda-Garcia et al., 2020); stakeholder collaboration and congruence in stakeholders' objectives (Perkins et al., 2020), and a country's overall image (Cai, 2002).

Benefits of Destination Branding

Destination branding offers several advantages. They include creating a sustainable competitive advantage that emphasizes exclusiveness (Castañeda-Garcia et al., 2020; WTO & ETC, 2009), building a positive image and a strong positioning (Cai, 2002), differentiating products (Park & Petrick, as cited in Séraphin et al., 2019), positively influencing destination choice (Blain et al., as cited in Castañeda-Garcia et al., 2020), and minimizing search costs (Erdem & Swait, as cited in Almeyda-Ibañez & George, 2017). Moreover, a brand provides a guarantee and protection (Besanko et al., as cited in Moilanen & Rainisto, 2009), extends goodwill (Murphy, as cited in Moilanen & Rainisto, 2009), and helps increase business turnover (Broniarczyk & Alba as cited in Moilanen & Rainisto, 2009).

Limitations

Destination brands have limited longevity, with a median 'age' of about five years (WTO & ETC, 2009). Less than 29% of destination marketing organizations used their brands for at least ten years; about 12% of brands lasted for only a year before being changed (WTO & ETC, 2009). Tourism and brands represent just two points of the competitive identity hexagon (Anholt, as cited in WTO & ETC, 2009). They may not be sufficient to overcome deficiencies in the other variables that comprise a country's competitiveness.

Studies on Destination Branding of the Philippines

The amount of published peer-reviewed research that mentions the Philippines' destination branding strategy is still very limited. In one study, using content analysis, Galí et al. (2016) examined the slogans of 150 destinations in the top 50 tourist destinations, where the Philippines was ranked 50th based on tourist arrivals and tourist expenditures. In that study, Gali et al. found that, unlike the Philippines, slogans were not used in 30 countries, which could have strengthened those countries' destination branding strategies. Another study examined the country's destination branding from a political lens. The Philippines Department of Tourism (DOT) slogan, 'Where Asia Wears a Smile', was used to defuse criticism of life under the New Society or the Philippine Style of Martial Law. The tourism campaign was made to depict a picture of peaceful and contented people (Richter, 1989). Villegas (2017) noted that the frequent changes in slogans, with different colour palettes, logos, and word lengths had made recall difficult for both domestic and international tourists. The slogans have failed to communicate impactful positioning for the Philippines.

Destination Branding Slogans and Their Impact on the Philippine Tourist Arrivals

This section presents the destination branding slogans of the country from 1973 to the present. The country never had an official tourism slogan before 1973 because the country only began to actively promote tourism in that year, after the DOT was formed. Prior to that, the Philippines was given the nickname, 'Pearl of the Orient Seas' or *Perla del Mar de Oriente*, which was first used in 1751. This term somehow stuck after the country's national hero, Jose Rizal, used the term in his last poem, 'Mi Ultimo Adios', before he died in 1896 ('Why Philippines is Called', 2012). The commentaries are from the authors' own recollections, as there are no longer any published materials on the slogans. Under the leadership of Secretary Jose Aspiras, the Department of Tourism promoted the country, using the following slogans:

Where Asia Wears a Smile
Philippines: An Island to Remember
The Philippines: A Nation Reborn
A Nation in Motion

The intention of the first slogan was to project a friendly and happy country, to counter the negative connotation of Martial Law, with the government pointing out that Philippine-style martial is different in that you cannot see any soldiers roaming the streets. The second slogan is misleading because the country is not one island but an archipelago. It also does not project any USP or brand identity that will make it stand out from the competition. The third and fourth slogans were meant to project a country that is modernizing and has found renewed pride. These are examples of politically motivated campaigns depicting a country that is undergoing fast transformation due to the massive infrastructure and facilities build up. The construction of a new airport, several five-star hotels, and convention venues was to support the influx of and stay of tourists, and the nascent meetings industry. The DOT's marketing efforts were criticized by many as being fragmented and disjointed, but inbound tourist arrivals still grew by an average of 14.5% from 1973 to 1985, the last full-year term of Aspiras.

After Marcos was toppled from power, the government under Corazon Aquino launched the new destination brand, 'Fiesta Islands'. These two words were supposed to capture the essence of the Philippines, first, because the country has a vibrant fiesta or festival culture, with every village having its own special festival in honour of a patron saint or to celebrate a bountiful harvest. The word, *islands*, is again an apt descriptor for the country, it being an archipelago. This slogan was implemented from 1986 to 1992. Even though the DOT went through four leadership changes under three secretaries during this period, it maintained this slogan. The DOT retained the word, *islands*, in the next round of promotional campaign from 1992 to 1998 during the incumbency of President

Fidel V. Ramos. During his term, the DOT had four secretaries and four different slogans as well.

Islands Philippines
Our Islands Have It
Best of the Islands
Musical Islands Philippines

These slogans, while offering some degree of identity to the country as islands, did not provide sufficient differentiation. In spite of not being in conformity with theoretical branding guidelines, the Philippines posted an average increase of 11% per year from 1993 to 1998. Joseph Estrada's presidency, which started in June 1998, ended abruptly after serving for just two-and-a-half years. Then DOT Secretary Gemma Araneta adopted the slogan, 'Rediscovery Philippines', in 2001. The slogan was used to sell eight specific destinations in the country, instead of the former 21 priority destinations. Tourist arrivals went down by -10% in 2001. This was not entirely because the slogan was weak but it could have been caused by the Asian financial crisis.

The DOT adopted the WOW Philippines slogan from 2002 to 2008 during the terms of President Gloria Arroyo (2001–2010). The campaign was the brainchild of Secretary Richard Gordon. He thought the catchy slogan could offset the negative publicity arising from kidnappings by the Abu Sayyaf and the political instability in Manila (Bellman, 2001). WOW stood for 'world of wonders' and had several sub-slogans as shown below. The sub-slogans were meant to promote specific tourism-related products.

WOW Philippines: World of Wonders
Sub-slogans
Watch Our Whales
Wild Over Water
Walk Our Walls
Warm Over Winter
Wacko Over Wildlife
Wear Our Wares

Wonderfully Original Waterways
World of Weddings
Wander Our Wonders
Workers of Worth
Wild Over Wheels

After Gordon, DOT Secretary Roberto Pagdanganan changed the slogan to *Pilipinas Kay Ganda* (Philippines So Beautiful) in 2010. However, the campaign was pulled out after netizens bashed the new logo as a copycat of a previous Poland tourism campaign 'Polska'. Surprisingly, during the two-year period from 2010 to 2011, tourist arrivals grew by an average of 14% per year. In 2012, DOT Secretary Ramon Jimenez, Jr. launched 'It's more fun in the Philippines' campaign. The campaign involved crowdsourcing and tapping social media users to post photos with the title of the caption. In support of this campaign, the DOT launched events on the website—www.experiencephilippines.org—and social network sites. The new slogan was accompanied by three sample memes for the public to model and spread through the Internet (Bosangit, as cited in Valdez et al., 2017). The campaign generated as many as 12,000 memes in less than a week, even with celebrities abroad participating in the campaign (Bosangit, as cited in Valdez et al., 2017). Considered as the most successful Philippine tourism campaign so far, it even won the third spot in the Warc 100, an annual ranking of the world's best marketing campaigns (Clift, 2014).

The slogan generated a lot of memes on social media. People posted on Facebook their own versions of why it is more fun in the Philippines. In terms of stakeholder engagement, 'it's more fun' was surely a winner. The campaign has a version for the domestic market, '#1for fun', but most people were not aware of it and simply used the version for the international market. During the years the slogan was used (2012–2016), the average annual increase in inbound arrivals was 9%. The words, 'It's more fun in the', were written on the first line and the word, 'Philippines', on the second line, all in Harabara font. The first line is written in dark blue, and the word, 'Philippines', was in shades of blue (Phil), yellow (i), orange (pp), and red (ines). The word Philippines is about 10% larger than the first line. On the right side of the slogan was the logo, an abstracted map of the Philippines, which the designer rendered like a banig or native mat. The copyright symbol is attached on the outside upper right corner of the mat element.

By 2016, Wanda Teo took over the reins of the DOT and proceeded to have the campaign changed to 'Experience the Philippines' and 'When you're with Filipinos, you're with family'. The secretary was quoted as saying that the new campaign would cost about US$13 million (ABS-CBN Investigative and Research Group, 2016). The DOT usually hires global advertising agencies to create the tourism campaigns, such as McCann Erickson in 2017 (De La Cruz, 2017) and BBDO in 2019 (Lopez, 2019). The move was prompted by Secretary Teo's statement that the ITMFITP had not been that effective in raising tourist arrivals in spite of its popularity ('How it's More Fun', 2016). She cited a Nielsen survey (Alpad, 2016) which showed that although 65% of respondents in Europe liked the 'It's more fun' campaign, only 26% had any visit intention. In North America, 72% said they liked the slogan, but only 45% had any plan to visit. She also said that it is customary to change the slogan when a new secretary is appointed (Alpad, 2016). However, the DOT was bombarded by complaints on social media and even some lawmakers who questioned the rationale for replacing a very successful slogan. For example, the then Senator J.V. Ejercito, who chaired the Committee on Tourism, maintained that tourism campaigns should not be revamped frequently because they cost taxpayers' money, noting that IMFITP was appropriate and had built a strong recall among foreign tourists ('Do not change…', 2017). Likewise, the president of a major tour operators association appealed to Teo to rethink the decision, since it would be difficult to introduce another slogan to the international market ('More fun in the Philippines', 2016). With Teo's resignation in May 2018 (Tan & de Guzman, 2018), the DOT would eventually relaunch the IMFITP slogan the following year.

As mentioned, the DOT retained the IMFITP campaign but with some revisions. This time, the words were rendered in Barabara font, with all letters in a capital case. Barabara font was designed by Filipino advertising art director, Michelle Co (Lopez, 2019), and was inspired by the style of lettering on Philippine jeepney signage (Rey, 2019). The first line contains the phrase 'It's more fun in the', and the second line contains the word 'Philippines'. The first line is now written in a lighter shade of blue, while the Philippines is still in rainbow colours, but with the addition of green shades. The word *Philippines* is about twice as large as the first line. The map logo disappeared. In its place are six diagonally oriented pieces of mat straws. Three straws are predominantly blue, with the other three in red. The logo is placed on the left side of the phrase. The copyright symbol is located at the bottom corner of the letter 's' of the Philippines. An illustration of this logo can be found using this link: http://beta.tourism.gov.ph/.

The new design is 'rooted in traditional motifs and colours'. The new logo was patterned after local woven textiles and represents 'a fresh graphic identity system' to symbolize DOT's fun, contemporary, and more sustainable offerings (Rocamora, 2019). The focus on sustainable tourism appears to be in line with the brand activism concept covered in this chapter's literature review. Using the hashtag, #ItsMoreFunIn ThePhilippines, the logo was shared nearly 4.6 million times. The DOT was able to crowdsource photos and videos from Instagram users for its advertising materials (Rey, 2019). Photos of most of the aforementioned Philippine destination brand slogans and logos may be viewed in the feature article by De La Cruz (2017).

Over the past 47 years, the DOT has used 14 destination brand slogans. That gives each slogan an average of 3.35 years. This is significantly shorter than Amazing Thailand, which was launched in 1998 ('TAT promotes "Amazing Thailand"', 1997) (22 years), and Malaysia Truly Asia, which has been used since 1999 (Kaur, 2019) (21 years). Because of their shorter lifespans, the Philippine slogans have failed to gain traction (Villegas, 2017). Compared to the three countries, the Philippines has a much lower number of foreign tourist arrivals in 2019, 8 million compared to Thailand's 40 million and Malaysia's 26.1 million. However, Incredible India, which was launched in 2002 (18 years), may not have produced spectacular arrival figures (11 million in 2019), but looking back to 2003, the first after the rollout of the campaign shows an impressive 14.3% surge in arrivals compared to -6% growth rate in 2002.

Back to Philippine tourism data, we look at the following year of the start of the campaigns. In 1974, after 'Wear Asia Wears a Smile', tourist arrivals jumped by 69%. In 1990, or a year after the Fiesta Islands' launch, tourist arrivals actually dipped by 14%. Islands Philippines generated an increase of 19% in 1993. After the launch of 'It's more fun in the Philippines', the country's tourist arrivals grew by 10% in 2013. However, in spite of the lack of originality of *Pilipinas Kay Ganda*, arrival figures increased by 17% in 2010 and 10% in 2011. During the period from 1973 to 1985, when the DOT had four different slogans, the average growth rate of tourist arrivals was 14.5%. From 1989 to 1992, when the DOT used just a single slogan, inbound travel increased by an average of 3.5%. From 1992 to 1998, the average annual growth rate was 12.6%, from 2003 to 2010, 8.1%.

The DOT also faced allegations of plagiarizing the tourism slogans of other countries. An online news article by Makabenta (2017) pointed out that Pilipinas Kay Ganda was copied from a Polish tourism campaign, 'It's more fun in the Philippines' (2012–2017) was similar to 'It's more fun in Switzerland'. Moreover, a 60-second television commercial under the Experience the Philippines (2017) campaign, featuring a blind Japanese tourist who settled in the Philippines and enjoyed the country's destinations, was of the same concept as one South African tourist campaign (Tables 11.2 and 11.3).

Based on the Philippine experience, it can be inferred that destination branding slogans have no direct connection to tourism performance as measured by the growth rates in tourist arrivals. The following observations could be made:

First, longer implementation of a destination brand slogan will not result in better competitive positioning. In the case of Incredible India!, tourist arrivals increased over the years since its implementation, but arrivals remain much lower in comparison to Malaysia and Thailand. In the Philippines, the Fiesta Islands campaign produced less than 50% of the tourist arrivals than the 'It's more fun' campaign even though the implementation period was almost the same.

Second, consistency in messaging was also not a factor in tourism performance. During the years, 1973 to 1985, when the DOT utilized four unrelated slogans, tourist arrivals grew by an average of 14.5%, which was much higher than the 1989 to 1982 period when the DOT used a single-destination-brand slogan. The former was also significantly

Table 11.2 Inbound tourist arrivals, 1972–2020

Year	Arrivals (in millions)	Growth rate
1972	0.17	15
1973	0.24	46
1974	0.41	69
1975	0.50	22
1976	0.62	22
1977	0.73	19
1978	0.86	18
1979	0.96	13
1980	1.01	4
1981	0.94	-7
1982	0.89	-5
1983	0.86	-3
1984	0.82	-5
1985	0.77	-5
1986	0.78	1
1987	0.80	2
1988	1.04	31
1989	1.19	14
1990	1.02	-14
1991	0.95	-7
1992	1.15	21
1993	1.37	19
1994	1.57	15
1995	1.76	12
1996	2.05	16
1997	2.22	8
1998	2.15	-3
1999	2.17	1
2000	1.99	-8
2001	1.80	-10
2002	1.93	8
2003	1.91	-1
2004	2.24	17
2005	2.62	17
2006	2.69	2
2007	3.09	15
2008	3.14	2
2009	3.02	-4
2010	3.52	17
2011	3.92	11
2012	4.27	9
2013	4.68	10

(*continued*)

Table 11.2 (continued)

Year	Arrivals (in millions)	Growth rate
2014	4.83	3
2015	5.36	11
2016	5.97	11
2017	6.62	11
2018	7.17	8
2019	8.26	15
2020	1.48	-82

Sources: Department of Tourism Statistical Reports & Cruz (2008), Department of Tourism (Visitor arrivals, 2008–2020); ASEAN Tourism Database (for years 1972–2007) in Cruz (2008)

higher than the Wow Philippines, and It's more fun in the Philippines campaigns. The only period when the tourist arrivals enjoyed double-digit growth was from 1992 to 1998 when four slogans consistently used the word, *Islands*. Surprisingly, in 2010 when the much-maligned Pilipinas Kay Ganda campaign was in effect, tourist arrivals actually grew by 17%, the highest tourism performance.

Third, stakeholder support and collaboration also did not influence arrival figures. For comparison, refer again to the tourist arrivals during *Pilipinas Kay Ganda* campaign period (weak stakeholder support), and the 'It's more fun in the Philippines' campaigns versions 1 and 2 (very strong stakeholder support and collaboration). Tourist arrivals during the implementation of IMFITP grew only by 8.8%, or just half of the *Pilipinas Kay Ganda* arrival figures. Version 2 failed miserably if tourist arrivals would be the sole basis.

The data show that a destination brand slogan is not the only determinant of tourism performance. There is a myriad of other variables that influence the trajectory of tourist arrivals that are not readily apparent from the numerical data. For example, the successive decline in arrivals in the early 1980s was fuelled by political unrest that was fanned by the assassination of a major political rival of President Marcos. Coup attempts against President Corazon Aquino, chronic power outages, a major earthquake on the main island of Luzon, and the volcanic eruption of Mt. Pinatubo also led to drastic decreases in tourist arrivals in 1990 and 1991.

Table 11.3 Philippines' destination branding slogans (1973–2020)

Inclusive Years	Destination Slogans	Average Annual Growth Rate in Tourist arrivals
1973–1985 (13 years)	Where Asia Wears a Smile Philippines: An Island to Remember The Philippines: A Nation Reborn A Nation in Motion	14.5%
1989–1992 (4 years)	Fiesta Islands Philippines	3.5%
1992–1998 (7 years)	Islands Philippines Our Islands Have It Best of the Islands Musical Islands Philippines	12.6%
2001 (1 year)	Rediscovery Philippines	-10%
2003–2010 (8 years)	WOW Philippines: World of Wonders Sub-slogans Watch Our Whales Wild Over Water Walk Our Walls Warm Over Winter Wacko Over Wildlife Wear Our Wares Wonderfully Original Waterways World of Weddings Wander Our Wonders Workers of Worth, Wild Over Wheels	8.1%
2010 (1 year)	"Pilipinas Kay Ganda," ["Philippines So Beautiful"]	17%
2012–2016 (5 years)	It's more fun in the Philippines Version 1	8.8%
2017–2018 (1 year)	Experience the Philippines	9.5%
2019–2020 (2 years)	It's more fun in the Philippines version 2	-82%

Sources: ABS-CBN Investigative and Research Group (2016), Cruz (2008), De La Cruz (2017) and Lopez (2019)

The Presidency of Fidel Ramos saw a period of double-digit growth in tourist arrivals, largely aided by the resolution of the power outages and peace agreements with leftist and Muslim rebel groups. However, the Asian financial crisis hit the country's tourist arrivals in 1998. The dismal tourism performance during 2000 and 2001 could be largely attributed to the SARS pandemic and the political turmoil that led to the ousting of the then president Estrada. Natural disasters of major proportions like Typhoon Ondoy in 2009 and the COVID-19 pandemic of 2020 have a far stronger impact on a country's tourist arrivals that destination brand slogans could not mitigate.

Conclusion

Progress in scholarship on destination branding has accelerated in the past decade both in quantity and in quality. While much of the conceptual foundations in destination branding was established in the 1990s, there have been fresh studies that advance the theory on destination branding and provide empirical support for such insights. Unfortunately, research on the topic is not that popular in the developing world, including the Philippines. Given the significant role that tourism plays in many developing economies, destination branding is one area of study that tourism academics in the developing states could explore further.

The approach to destination branding in the Philippines by the DOT is more a matter of tradition than one driven by science. As is noted, it was customary for an incoming secretary to embark on a new campaign. So far, there were 20 appointments to the post, involving 18 individuals. Since the DOT spends tens of millions of US dollars for each campaign, frequent changes in the destination slogans represent wasteful spending of taxpayers' money in a developing country like the Philippines.

The data also demonstrate that there is no direct link to the destination brand slogans with tourism performance, with seemingly poorly conceived campaigns like *Pilipinas Kay Ganda* and Experience the Philippines producing better results in terms of arrivals than the highly regarded campaigns such as 'Wow Philippines', and 'It's More Fun in the Philippines' versions 1 and 2. Thus, it could be inferred that destination branding cannot be isolated from the larger sociopolitical, environmental, and economic realities happening in the national and global settings.

Going forward, it would be best for NTOs in the emerging economies to establish a systematic assessment system for measuring destination brand slogan effectiveness. It is important to adopt metrics such as the number of tourist receipt for every dollar spent on a destination brand campaign and to conduct regular surveys at the gateway ports that investigate whether there is a direct link between the destination brand and tourists' decision to come to the country. This will also help emerging economies to design their slogans based on research-based frameworks.

References

Aaker, D., & Joachimsthaler, E. (2000). The brand relationship spectrum: The key to the brand architecture challenge. *California Management Review, 42*(4), 8–23. https://doi.org/10.1177/000812560004200401

Adeyinka-Ojo, S., & Nair, V. (2016, May 11–13). Destination branding framework in tourism research: A state-of-the-art analysis [Conference paper]. 14th APacCHRIE Conference, Bangkok, Thailand. https://www.researchgate.net/publication/303330884_Destination_Branding_Framework_in_Tourism_Research_A_State-of-the-Art_Analysis

Almeyda-Ibañez, M., & George, B. P. (2017). The evolution of destination branding: A review of branding literature in tourism. *Journal of Tourism, Heritage & Services Marketing, 3*(1), 9–17. https://doi.org/10.5281/zenodo.401370

Alpad, C. (2016, October 8). Still fun, but change is coming. *The Sunday Times Magazine.* https://www.manilatimes.net/2016/10/08/weekly/the-sunday-times/still-fun-change-coming/290120/

Bae, S., Jung, T. H., Moorhouse, N., Suh, M., & Kwon, O. (2020). The influence of mixed reality on satisfaction and brand loyalty in cultural heritage attractions: A brand equity perspective. *Sustainability, 12*(7), 2956. https://doi.org/10.3390/su12072956

Bellman, E. (2001, August 7). Philippines struggles to find catchy slogan for tourism. *The Wall Street Journal.* https://www.wsj.com/articles/SB997204075671312135#:~:text=The%20Philippines%20has%20run%20the,for%20the%20past%20five%20years

Bianchi, C., Pike, S., & Lings, I. (2014). Investigating attitudes towards three South American destinations in an emerging long haul market using a model of consumer-based brand equity (cbbe). *Tourism Management, 42*(2), 15–23. https://doi.org/10.1016/j.tourman.2013.11.014

Boo, S., Busser, J., & Baloglu, S. (2009). A model of customer-based brand equity and its application to multiple destinations. *Tourism Management, 30*(2009), 219–231. https://doi.org/10.1016/j.tourman.2008.06.003

Cai, L. A. (2002). Cooperative branding for rural destinations. *Annals of Tourism Research, 29*(3), 720–742. https://doi.org/10.1016/S0160-7383(01)00080-9

Castañeda-Garcia, J. A., Frías-Jamilena, D. M., Del Barrio-García, S., & Rodríguez-Molina, M. A. (2020). The effect of message consistency and destination-positioning brand strategy type on consumer-based destination brand equity. *Journal of Travel Research, 59*(8), 1447–1463. https://doi.org/10.1177/0047287519881506

Chigora, F., & Zvavahera, P. (2015). Awareness completes brand loyalty: Reality of Zimbabwe tourism destination. *Business and Management Horizons, 3*(2), 60–69. https://doi.org/10.5296/bmh.v3i2.8547

Clift, J. (2014, November 7). Warc 100: Why it's more fun in the Philippines. *Warc.* https://www.warc.com/newsandopinion/opinion/warc-100-why-its-more-fun-in-the-philippines/1879

Cruz, R. G. (2008). *Principles of travel and tourism.* Unpublished manuscript.

De La Cruz, C. I. (2017, June 9). Do you remember these Department of Tourism slogans? *Spot.* https://www.spot.ph/newsfeatures/the-latest-news-features/70411/department-of-tourism-slogans-a00171-20170609

Do not change 'It's more fun in the Philippines', Ejercito tells DOT. (2017, June 13). *ABS-CBN News.* https://news.abs-cbn.com/news/06/13/17/do-not-change-its-more-fun-in-the-philippines-ejercito-tells-dot

Dyankov, T., Krastev, V., & Yancheva, K. (2018). Evaluation of brand identity of Bulgaria as a tourist destination. *Tourism: An International Interdisciplinary Journal, 66*(1), 19–34. https://hrcak.srce.hr/197380

Galí, N., Camprubi, R., & Donaire, J. A. (2016). Analysing tourism slogans in top tourism destinations. *Journal of Destination Marketing & Management, 6*(3), 243–251. https://doi.org/10.1016/j.jdmm.2016.04.004

Garcia, J. A., Gómez, M., & Molina, A. (2012). A destination-branding model: An empirical analysis based on stakeholders. *Tourism Management, 33*(2012), 646–661. https://doi.org/10.1016/j.tourman.2011.07.006

Hanna, S., & Rowley, J. (2011). Towards a strategic place brand-management model. *Journal of Marketing Management, 27*(5–6), 458–476. https://doi.org/10.1080/02672571003683797

Harish, R. (2010). Brand architecture in tourism branding: The way forward for India. *Journal of Indian Business Research, 2*(3), 153–165. https://doi.org/10.1108/17554191011069442

How It's more fun in PH slogan fared from 2012 to 2016. (2016, November 24). ABS-CBN Investigative and Research Group. https://news.abs-cbn.com/focus/11/24/16/how-its-more-fun-in-ph-slogan-fared-from-2012-to-2016

Hsu, C., & Cai, L. A. (2009, July 31). Brand knowledge, trust and loyalty—A conceptual model of destination branding. International CHRIE Conference-Refereed Track. https://scholarworks.umass.edu/cgi/viewcontent.cgi?article=1110&context=refereed

Kaur, K. (2019, July 23). 1990 to 2020 the history of Visit Malaysia logos. *The Rakyat Post*. https://www.therakyatpost.com/2019/07/23/1990-to-2020-the-history-of-visit-malaysia-logos/

Kim, H., & Lee, T. J. (2018). Brand equity of a tourist destination. *Sustainability, 10*(2), 431. https://doi.org/10.3390/su10020431

Kotler, P. (1999). *Kotler on marketing: How to create, win and dominate markets*. Free Press. https://www.researchgate.net/publication/265069529_Kotler_on_Marketing_How_to_Create_Win_and_Dominate_Markets

Kotler, P. (2016). Branding: From purpose to beneficence. *The Marketing Journal*. https://www.marketingjournal.org/brand-purpose-to-beneficence-philip-kotler/

Kotler, P., & Sarkar, C. (2017). Finally brand activism! *The Marketing Journal*. https://www.marketingjournal.org/finally-brand-activism-philip-kotler-and-christian-sarkar/

Listiana, E. (2015). Country of origin image and its impact on brand Association, perceived quality and brand loyalty. *Journal of Marketing and Consumer Research, 10*(2015), 19–29. https://citeseerx.ist.psu.edu/viewdoc/download?doi=10.1.1.877.7916&rep=rep1&type=pdf

Lopez, E. C. (2019, February 19). Meet the 25-year-old designer behind the DOT's 'Barabara' font. *Esquire*. https://www.esquiremag.ph/culture/lifestyle/dot-barabara-font-designer-a00290-20190219

Lund, N. F., Cohen, S. A., & Scarles, C. (2018). The power of social media storytelling in destination branding. *Journal of Destination Marketing & Management, 8*, 271–280. https://doi.org/10.1016/j.jdmm.2017.05.003

Makabenta, Y. (2017, June 20). Three times, DOT copied tourism campaigns of other states. *The Manila Times*. https://www.manilatimes.net/2017/06/20/

opinion/columnists/topanalysis/three-times-dot-copied-tourism-campaigns-states/333679/

Mlambo, S. S., & Ezeuduji, I. O. (2020). South Africa's Kwazulu-Natal tourism destination brand essence and competitiveness: Tourists' perspectives. *GeoJournal of Tourism and Geosites, 32*(4), 1195–1201. https://doi.org/10.30892/gtg.32401-557

Moilanen, T., & Rainisto, S. (2009). *How to brand nations, cities and destinations: A planning book for place branding* (1st ed.). Palgrave Macmillan. https://www.academia.edu/10219180/How_to_brand_nations

'More fun in the Ph' slogan soon to be history. (2016, September 5). *Coconuts Manila.* https://coconuts.co/manila/news/more-fun-ph-slogan-soon-be-history/

Pan, S. (2019). Tourism slogans—Towards a conceptual framework. *Tourism Management, 72*(2019), 180–191. https://doi.org/10.1016/j.tourman.2018.11.023

Perkins, R., Khoo-Lattimore, C., & Arcodia, C. (2020). Understanding the contribution of stakeholder collaboration towards regional destination branding: A systematic narrative literature review. *Journal of Hospitality and Tourism Management, 43*(2020), 250–258. https://doi.org/10.1016/j.jhtm.2020.04.008

Pike, S. D. (2004). Destination brand positioning slogans—Towards the development of a set of accountability criteria. *Acta Turistica, 16*(2), 102–124. http://eprints.qut.edu.au/

Qu, Y., Cao, L., & Xu, F. (2021). Design of an attention-grabbing destination slogan using the attenuation model. *Journal of Destination Marketing & Management, 19*(2021) https://doi.org/10.1016/j.jdmm.2020.100415

Rey, A. (2019, February 18). 'It's more fun in the Philippines' gets new font. *Rappler.* https://www.rappler.com/business/its-more-fun-philippines-new-font

Richter, L. (1989). The Philippines: The politicization of tourism. In L. A. Richter (Ed.), *The politics of tourism in Asia* (pp. 57–91). University of Hawaii Press. https://doi.org/10.2307/j.ctv9zcjr9.7

Rocamora, J. (2019, 18 February), PH launches refreshed tourism slogan. *Philippine News Agency.* https://www.pna.gov.ph/articles/1062242

Ruiz-Real, J. L., Uribe-Toril, J., & Gázquez-Abad, J. C. (2020). Destination branding: Opportunities and new challenges. *Journal of Destination Marketing & Management, 17*(2020), 100453. https://doi.org/10.1016/j.jdmm.2020.100453

Saqib, N. (2020). Positioning—A literature review, *PSU Research Review*, Vol. ahead-of-print No. ahead-of-print. https://doi.org/10.1108/PRR-06-2019-0016

Séraphin, H., Zaman, M., Olver, S., Bourliataux-Lajoinie, S., & Dosquet, F. (2019). Destination branding and overtourism. *Journal of Hospitality and Tourism Management, 38*(2019), 1–4. https://doi.org/10.1016/j.jhtm.2018.11.003

Tan, L., & de Guzman, C. (2018, May 8). Tourism Secretary Wanda Tulfo-Teo resigns 'out of delicadeza. *CNN Philippines*. https://cnnphilippines.com/news/2018/05/08/Duterte-Tourism-Secretary-Wanda-Teo-resign-Tulfo.html

TAT promotes "Amazing Thailand Campaign" 1998–1999 through international phonecards. (1997, July 14). https://www.ryt9.com/en/prg/1690

Tran, V. T., Nguyen, N. P., Tran, P. T. K., Tran, T. N., & Huynh, T. T. P. (2019). Brand equity in a tourism destination: A case study of domestic tourists in Hoi An City, Vietnam. *Tourism Review, 74*(3), 704–720. https://doi.org/10.1108/TR-08-2017-0130

Valdez, P. N. M., Tupas, R., & Tan, N. C. (2017). "It's more fun in the Philippines": Resemiotizing and commodifying the local in tourism discourse. *Discourse, Context & Media, 20*, 132–145. https://doi.org/10.1016/j.dcm.2017.09.002

Villegas, V. (2017). Building a successful tourism brand for the Philippines and the struggle with destination image and tourism slogans. *Luz Y Saber, 11*(1), 62–79. https://research-manila.letran.edu.ph/read/111

Visitor arrivals to the Philippines. (2008–2020). Department of Tourism. http://tourism.gov.ph/tourism_dem_sup_pub.aspx

Why Philippines is called Pearl of the Orient Seas. (2012, January 11). Come Visit Fun Philippines. http://www.strawhatjp.com/why-philippines-is-called-pearl-of-the-orient-seas/

World Tourism Organization & European Travel Commission. (2009). *Handbook on tourism destination branding*.

World Travel and Tourism Council. (2020). Travel and tourism global economic impact and trends. https://wttc.org/Portals/0/Documents/Reports/2020/Global%20Economic%20Impact%20Trends%202020.pdf?ver=2021-02-25-183118-360

12

Opportunities and Challenges of Globalisation for ASEAN Destinations Through the One Belt One Road Initiative

Agus Riyadi, Mohd Azlan Abu Bakar, and Candra Hidayat

Introduction

One of the rapidly developing regional cooperation bodies, the Association of Southeast Asian Nations (ASEAN), has shown steady progress in economic and social development, with many positive views on the upcoming cooperation on many fronts. The recent decision by ASEAN countries to submit a joint bid to stage the 2034 World Cup has caught many by surprise. During the 34th ASEAN Summit, the announcement

A. Riyadi (✉) • C. Hidayat
Trisakti School of Tourism, Jakarta, Indonesia
e-mail: agus.riyadi@stptrisakti.ac.id; candra.hidayat@stptrisakti.ac.id

M. A. A. Bakar
Faculty of Management, PICOMS International University College, Kuala Lumpur, Malaysia
e-mail: azlan@picoms.edu.my

I. Mensah et al. (eds.), *Marketing Tourist Destinations in Emerging Economies*, Palgrave Studies of Marketing in Emerging Economies,
https://doi.org/10.1007/978-3-030-83711-2_12

demonstrated that ASEAN discussion on regional cooperation encompassed various regional issues, including the most popular sports in the world, football. ASEAN seeks to promote intergovernmental cooperation in economic, military, security, educational, and other common interests towards the development of the ASEAN region.

During the early post-independence era, agriculture and mining were the top two industries in the ASEAN. The crops and raw materials were then exported to the Western countries for manufacturing purposes. Realising the potential of intraregional economic collaboration, ASEAN has formulated a policy known as 'Towards One Economic Community' that aims at achieving four pillars, specifically (a) single market and production base, (b) competitive economic region, (c) equitable economic growth, and (c) integration into the global economy (ASEAN, 2017). This formula has resulted in a more connected community where total tourist arrivals in ASEAN have increased, with half of the arrivals being intra-ASEAN.

Since 1967, ASEAN has initiated Frameworks for Regional Cooperation, which aims at different outcomes, namely science and technology, private sector engagement, energy, tourism, transport, finance, and agriculture. Tourism was among the first measures taken in making ASEAN a world marketplace. It was formulated in 1976 by forming a sub-committee on tourism that promotes travel within the region. This was made easier with a structured nation-building effort to revolutionise infrastructure as well as basic accommodation and travel facilities like airports, hotels, immigration bureau, and banking systems.

Malaysia was among the first countries to promote inter-regional tourism with the foundation of the Tourist Development Corporation in 1972 under the Ministry of Trade and Industry (Tourism Malaysia, 2018). Malaysia was also among the first countries in ASEAN to have a structured tourism campaign with a 'Visit Malaysia 1990' campaign. Later, the Visit ASEAN Campaign (VAC) was launched to promote brand building for regional tourism. VAC aims to establish an aggressive promotional campaign in advertising ASEAN as a single destination. Nonetheless, there is no clear indication of the outcome of these efforts, especially the 'Visit ASEAN@50' golden jubilee (1967–2017) campaign (Scott, 2007).

In recent years, tourism has exploded in Indonesia, and it is now one of the country's primary sources of foreign currency earnings. As reported by the Organisation for Economic Co-operation and Development (OECD), tourism contributed US$37.4 billion to Indonesia's GDP or 4.1% of the country's total GDP and created 12.7 million jobs in the same year, accounting for 10.5% of total employment in 2017 (OECD, 2020). In the Philippines, government policies strongly impact tourism policy planning and growth, indicating a top-down approach, which means tourism policies and decisions are dictated by the government through a hierarchical and centrally controlled method (Capistrano & Notorio, 2020). Compared to other ASEAN countries, Thailand's tourism industry is highly significant, accounting for 16.6% of the country's GDP in 2015, which was higher than the global average of 9.8% (Ministry of Tourism & Sports Thailand, 2017). Tourism receipts have also been increasing at the rate of 15.6% per year since 2011, and the industry has employed over 4.2 million people, accounting for 11% of the total workforce in the country (Ministry of Tourism & Sports Thailand, 2017). For Singapore, the tourism industry recorded 3.3% growth in 2019, compared to the previous year, with 19.1 million arrivals and US$20.9 billion receipts (Heng, 2020).

Impact of Globalisation on Destinations

Intraregional collaboration is not a new phenomenon anymore because almost every region has its economic cooperation. The most potent intra-regional economic pacts, the European Union (EU) and North Atlantic Treaty Organisation (NATO), had promoted 'globalisation' as a new economic order for the world. The main idea of globalisation is 'free trade', where there are no barriers to the movement of goods and services. This is commonly regarded as 'the level playing field' market. With the advancement of information communications technologies (ICTs), the relationship between continents and countries has become more dynamic, where people, business, materials, and information would get connected easily. The resistance to globalisation was witnessed by developing countries led by Nelson Mandela at the 1999 World Economic Forum held at

Davos, where he argued the advantages of globalisation (World Economic Forum, 2013). According to Mandela, globalisation will only benefit the financiers, investors, traders, and speculators, but not low-income earners. According to Mohamad (2002), the challenge to attract multinational corporations was real, whereby for every 100 economic entities in the world, 51 are global corporations while the remaining are countries. The combined sales of the world's top 200 corporations exceed the combined GDPs of 182 nation states (Mohamad, 2002).

Based on the theory of relative advantage, free trade will encourage more foreign direct investment (FDI) in developing countries and create a labour-intensive market for the host country. Still, many arguments were directed to this assumption, especially when other economic variables were put into the equation, particularly productivity, competitiveness, domestic political situation, fiscal policy, and automation. In other words, there is a disparity between demand-led economic (employment) growth against supply constraints (lack of infrastructure, skilled labour, and government participation) (Lee & Vivarelli, 2006).

The ASEAN Economic Community (AEC) is an initiative of ASEAN countries to transform ASEAN into a sustainable and resilient economic region in the global economy (ASEAN, 2013). According to Song et al. (2018), the globalisation of the tourism sector has resulted in the fragmentation of the tourism production framework and the trans-nationalisation of ownership systems, marketing agreements, service outsourcing, and knowledge transmission. Agus and Indra (2018) added that globalisation enabled previously unimaginable products, such as slums and poverty, to become commodities and cultural heritage. The emergence of digital media, such as the Internet, has succeeded in shaping our current reality. For instance, 'The Jakarta Secret Tour' uses the Internet to spread the message of poverty around the world. In the case of 'The Jakarta Secret Tour', the process of raising consciousness about poverty is linked to the destination marketing concept (Agus & Indra, 2018). Furthermore, globalisation contributed to the development of the modern global transportation system, allowing remote destinations to be reached easily and quickly, as well as the establishment of facilities such as airports, hotels, and resorts. Governments gradually streamlined structured procedures to facilitate the processing of increasing numbers of visitors (Cohen, 2012).

The new sharing economy mechanisms which fully utilise technologies to streamline transactions and make the property more available to owners and users worldwide are successful business models in the globalised world (Kostková, 2020). Even in a borderless society, technology allows for cross-border collaboration and sharing platforms that connect owners of underutilised resources with people who want to access or use them. Compared to the traditional hotel and accommodation business model, globalisation creates shared values, lifestyles, and spending habits, resulting in a universal rent-seeking culture (Zhou et al., 2019). In today's world, corporate and economic decisions are no longer made with a myopic view of fulfilling local restrictions, but instead considering the influence of such actions on the global economy (Saint Akadiri et al., 2019). In essence, globalisation is progressively influencing how businesses are done, as enterprises now have access to global resources rather than relying on local resources, as was the case previously.

Globalisation has been a hot topic in destination research, with researchers debating the impact of globalisation on lifestyles, economic growth, and society's health and social well-being due to climate change (Balsalobre-Lorente et al., 2020). The global crisis brought on by the coronavirus pandemic has profoundly altered the worldwide tourism business. As a result, there was a need to re-evaluate how the tourism business did its typical operations. Perhaps, there will be a new generation of destination marketing solutions that focus on how virtual solutions and remote work may thrive in the age of digital technology and reduced global mobility (Kyrylov et al., 2020).

ASEAN Tourism in an Era of Globalisation

The West mooted the idea of globalisation at the end of the Cold War period. The free market will encourage mobility of goods (cheap raw materials), expertise (transfer of knowledge), capital (foreign direct investment), socio-economic development (improved income), and mobility (movement of people) in a single marketplace. From the literature review, lots of discussions were directed at the effect of globalisation towards 'openness' in the form of the economy, 'sharing' in the form of expertise, 'equitable' distribution of wealth, and 'free' movement of goods and people. Nonetheless, the effect of globalisation on the travel and tourism industry received considerably less attention during the 1990s. As mentioned by the United Nations (2004), regional development was the key to tourism progress in the globalisation era. ASEAN, for instance, has pledged to mutually cooperate in the tourism sector with the formation of a sub-committee composed of ministers and experts.

As early as the 1960s, there was a discussion to develop four major transport corridors to connect Asia and Europe through the Northern Corridor (Korean Peninsula to Europe), Southern Corridor (South-East Asia [SEA] to Europe), Indo-China and ASEAN sub-regions, and North-South Corridor (Northern Europe to the Persian Gulf). In addition, the Pacific Islands Forum 1998 action plan has outlined cost-effective airspace management as one way to boost tourism. In the early days of globalisation, connectivity through air and land transportation, national security and sovereignty, as well as national and cultural identity became the subject of discussion, with less attention for duration and cost of travelling.

International tourism in the 1990s was dominated by European travellers who accounted for 60% of the international tourist arrivals. The Americas was the second largest segment (21%), followed by Asia and the Pacific (8%). Statistics showed that the number of tourist arrivals had increased by 38% from 435 million in 1990 to 674 million by the year 2000 (UNWTO, 2011). With an increment of more than 15% annually, the tourism industry has changed its status from a mere peripheral industry to be among the top three industries in ASEAN. Tourist arrivals correlated positively with tourism receipts, whereby there was a significant improvement by 375% from US$104 billion (1980) to US$495 billion (2000) (UNWTO, 2011). The positive performance of the tourism industry had prompted ASEAN countries to spend more on attracting international hotel brands into the country.

In the 1990s, tourist arrivals in ASEAN accounted for less than 7% of the total international tourist arrivals. Malaysia, Singapore, and Thailand were the key tourism markets in ASEAN, where they held more than 20% of tourist arrivals in the region. Malaysia is the key for intra-ASEAN tourism, with more than 40% of tourist arrivals. During the globalisation era, extra-ASEAN tourism contributed to about 60% of the tourist arrivals during the same period (ASEAN, 2013).

In the early globalisation phase throughout the 1990s, the top three extra-ASEAN markets came from Japan, Taiwan, and South Korea, while the rest came from the United States, the United Kingdom, and Australia (Table 12.1). ASEAN showed a steady increment in tourist arrivals despite inadequate facilities like smaller airports, lack of integration in public transportation, shortage of star-rated accommodation, and lack of coordination in the operation of tourist facilities. Despite the lack of a modern and integrated transportation system, international tourists had no problem visiting tourist spots with the popular conventional modes of public transportation like coach, minibus, taxi, and locally made vehicles like *Tuk-tuk* (Thailand) and *Angkot* (Indonesia). Furthermore, group travellers were more popular than fully independent travellers (FIT) during the 1990s. The breakthrough of the ASEAN tourism industry occurred in the mid-2000s, with the revolution in low-cost carriers, particularly AirAsia, Cebu Pacific, and Lion Air. Expensive airfares and limited travel options are a thing of the past in ASEAN, thanks to the emergence of low-cost carriers. ASEAN's low-cost carriers have reshaped the region's aviation landscape, bridging national borders and making travel more accessible to the masses (The ASEAN Post, 2021). ASEAN's seat size has more than doubled in the past decade, rising from just over 200 million seats in 2008 to nearly 530 million seats in 2018 (Centre for Aviation, 2019). In addition, ASEAN has consistently grown in the high

Table 12.1 Visitor arrivals to ASEAN by country of origin and year during 1995–1999

Country of destination	1995	1996	1997	1998	1999
ASIA					
Japan	3295	3664	3652	3070	3346
Taiwan	2141	2262	2031	1646	1747
South Korea	1120	1388	1267	596	1018
Rest of the World					
USA	1443	1441	1577	1637	1748
Australia	1084	1201	1457	1387	1602
United Kingdom	1032	1056	1114	1231	1304

NB. Figures are in a million
Source: ASEAN (2013)

single digits or low double digits, making it one of the fastest growing regions on the globe (Centre for Aviation, 2019).

In the mid-2000s, the demand for ASEAN tourism saw a changing pattern from the major Asia market. There was a drastic change in visitor arrivals in ASEAN, where China accounted for 40% of all Asian travellers. A giant leap was evident in 2011, with a 35.1% annual increment from Chinese visitors. It is also worth noting that India has accounted for 9% of the total Asian travellers. In this new environment, it is worth mentioning that China has to make a big wave in the ASEAN market. From the statistics published by ASEAN, the top three ASEAN markets for China travellers are Singapore, Thailand, and Vietnam (ASEAN, 2014).

The unexpected influx of China travellers into the ASEAN market is a result of the country's rapid economic progress, rising individual wealth, and softer travel restrictions (Tawil & Al Tamimi, 2013). Furthermore, the improved diplomatic relation with ASEAN has contributed to the strong relationship between China and ASEAN. As touted by Prime Minister Wen Jiabao, China considers itself a 'gentle and friendly elephant' to its smaller counterparts (Chen, 2006). China and ASEAN need each other for a trade alliance. China has economic power, while ASEAN has an abundance of untapped resources in land, people, and technology. The strong ties between ASEAN and China were the primary reason for massive capital funding to the ASEAN market by the Government of China under the 'One Belt One Road' initiative.

One Belt One Road Initiative (China and ASEAN)

A 'borderless' world required a comprehensive and integrated transportation system. ASEAN has successfully executed the 'ASEAN Open Skies Policy' since 2015 that intended to increase regional and domestic connectivity and enhance regional trade by allowing airlines from ASEAN countries to fly freely throughout the region under a single, unified air transport market (ASEAN Briefing, 2015). Fifty years before implementing the ASEAN Open Skies Policy, the idea of developing transport corridors was already in the pipeline, with four gazette routes that connect Europe and Asia through China as the primary access. The idea has not yet been implemented, but China has made a bold initiative under the 'New Economic Belt of the Silk Road' and the '21st Century Maritime Silk Road' famously called the 'One Belt One Road' initiative (BRI) in the Asian countries. This initiative, which will ensure connectivity through infrastructure development, covers 65 countries and involves over 60% of the world's population.

BRI is a grandmaster plan to (a) tap the land bank or reserved land of developing countries and (b) build the manufacturing industry and facilities by leveraging China's expertise. China has pledged more than US$5 trillion of funding to build landmarks, infrastructure, and facilities that include transport, energy, roads, bridges, gas pipelines, ports, railways, and power plants (Huang, 2017). For ASEAN, BRI will accelerate the ASEAN Master Plan for Connectivity (AMPC) and help to fund the mega transport projects that will boost the productivity and accessibility of its member countries. This initiative will increase the bilateral trade between China and ASEAN from US$472.2 billion in 2015 to US$1 trillion by 2020. The work in progress includes a new railway (high-speed rail—HSR) line from China to Singapore through Laos and Thailand. A 150 km HSR is also being constructed between two major cities in Indonesia (i.e. Jakarta and Bandung). Less developed ASEAN members, comprising Cambodia, Laos, and Myanmar, will see the development of the railway, power plant, and property that will change the economic outlook for those countries. Thus far, the 'New Economic Belt of the Silk Road' has been progressing quite well, but not the '21st Century Maritime Silk Road'. There are disputes over sovereignty in the South China Sea that led to a tense relationship between China and Vietnam. However, Malaysia and the Philippines have an agreement over maritime cooperation that will give China a territorial advantage against the United States. Another issue surrounding BRI is its execution, whereby the construction team, materials, and technology must come from China to ensure the big infrastructure projects are completed (Jetin, 2018). This condition will not help ASEAN to tackle the high unemployment rate, especially at the supervisory level.

Most of the initiatives taken under BRI concentrated on the manufacturing industry, with a lack of emphasis on the tourism industry. BRI will create a new environment for developing countries where transportation will no longer be an issue for travel. The duration of travel will be reduced, and the cost of travel will reduce as compared to other modes of transportation. The initial report noted that infrastructure, trading, and financing would get the most from BRI, with the food industry being among the top beneficiaries, through better product flow within the Asian communities. In other words, local products could be transported easily to China and vice versa. This situation could adversely affect local food establishments that sell food to Chinese tourists who visit South-East Asia (SEA), where due to the availability of Chinese food, demand for local food has declined. Gastronomic experiences of tourists constitute an emerging segment for both countries, where it contributed 10–15% to the food and beverage (F&B) segment. Adopting a single currency will be an interesting issue for Asian nations to overcome over-dependency on the US dollar in trading activities. However, it is not clear whether Asia could replicate the European model of introducing a single currency for the benefit of all trading partners.

BRI provides hope for ASEAN in five significant areas, namely (a) policy coordination, (b) connecting infrastructure, (c) unimpeded trade, (d) financial integration, and (e) people-to-people bonds (Lau, 2017). Policy coordination is a critical aspect of BRI because there is a risk of political instability affecting the execution of BRI. There should be a code of conduct to help avoid disputes resulting from the execution of the project. Fifty years since its inception, ASEAN has the experience to curb disputes over territorial ownership, especially in the South China Sea. Also, connecting infrastructure will bring alternative and perhaps cheaper cross-border options for people and goods. Connected and integrated infrastructure as in Europe will boost tourism and business activities within the vicinity of the transportation network. Apart from that, funding by China will help to resolve the issues of the inadequacy of public funding and accelerated project costs and avoid trade costs resulting from inadequate facilities and bureaucracy.

ASEAN can market the small- and medium-sized industry products to other parts of the world at a faster pace through multichannel transportation either by rail, sea, or air cargo. Besides that, financial integration can be executed through the regional banking system. There will be more mergers and acquisitions of financial institutions for investment and lending activities. Finally, BRI will create a large society that interacts with cultural and academic exchanges (Lau, 2017). People will have a chance to learn about other cultures through interaction with foreign employees that work for multinational organisations. This will also promote cultural interaction between students in the primary, secondary, and tertiary levels, resulting in a harmonious and tolerant community.

BRI and the Future Tourism Outlook of Asia

The tourism industry will reap the benefits of the 'One Belt One Road' (BRI) initiative. Firstly, BRI will raise the GDP of ASEAN by more than 5% (WTTC, 2016). Chinese travellers will increase further with the accessibility of the ASEAN markets through integrated rail transportation that links Southern China to Singapore via Laos, Cambodia, and Thailand. There have been several concerns on the ASEAN market about the infrastructure constraints, especially in Myanmar, Laos, the Philippines, Vietnam, and Cambodia (WTTC, 2016). However, this will be resolved with the high-speed rail (HSR) projects funded by the Government of China.

ASEAN countries must meet the demands of Chinese travellers for more personalised, authentic, and exclusive travel packages (ASEAN, 2012). Safety and security should no longer be an issue with the vast investment channelled into ICT and automation. Furthermore, BRI should be extended to investment in telecommunications and wireless facilities to allow cashless facilities. 'Borderless' should come in the form of a single Asian pass for ease of travelling. Reform in the banking and financial system is inevitable in a borderless world. A joint venture or merger and acquisition will provide convenience for utilising banking facilities during travel. The world is evolving into a 'cashless society', where transactions are completed without exchanging physical money. Lodging and food industries are the major industries that will benefit from the integrated online payment system.

Disruptive technology has changed the way people travel, whereby online travel agencies provide a one-stop centre for booking and reservation of rooms, airlines, and travel packages. With better accessibility, the sharing economy will be enhanced through point-to-point transfer services. People will no longer need a car for travelling. Mobile applications will play a more significant role in the BRI era. There is a high possibility of a spin-off effect for Grab Company (e-hailing services) to be expanded as 'Grab Travel' in the future. In addition, rapid urbanisation may occur in the new cities that are located along HSR stations. Urbanisation has both positive and negative impacts from a socio-economic standpoint. Though urbanisation will create more jobs for local people, it will also create social and environmental problems. ASEAN and China need to promote responsible tourism for their communities. As of now, BRI places tourism in a strategic role across 33 countries, with marketing and promotion, capacity building and destination management, and travel facilitation (Li et al., 2020).

Conclusion

BRI is an excellent globalisation initiative by China and ASEAN in realising a borderless world using an Asian approach. However, a strategic analysis of the tourism industry in the ASEAN market needs to be done to assess the real impacts of BRI on ASEAN and China. China can lead the initiative of making ASEAN + China the next important tourist destination by regularly engaging with experts and policymakers to assess the progress of BRI while discussing the potential threats of BRI to national and regional unity. ASEAN needs to develop more travel packages that can appeal to different traveller segments. In addition, renovation of rundown tourist facilities needs to be done right away before completing new state-of-the-art facilities. Also, the countries involved in the BRI project have different political and economic structures and different cultural backgrounds. There are numerous legal, political, and other threats. Finally, the only way for a destination to expand in size and power, reform, and evolve is through globalisation. Although the BRI initiative has the potential to create unprecedented opportunities like easy access to new investment and foreign talents in the coming decades, it also has the potential to create a growing threat of over-dependency to China in tourist arrivals and developing domestic tourism industry.

References

Agus, P., & Indra, K. (2018). The Indonesian slum tourism: Selling the other side of Jakarta to the world using destination marketing activity in the case of "Jakarta hidden tour". *E3S Web of Conferences, 73*, 08017. https://doi.org/10.1051/e3sconf/20187308017

ASEAN. (2012). *ASEAN tourism marketing strategy 2017–2020*. The ASEAN Secretariat.

ASEAN. (2013). *ASEAN statistical yearbook 2013*. The ASEAN Secretariat.

ASEAN. (2014). *ASEAN statistical yearbook 2013*. The ASEAN Secretariat.

ASEAN. (2017). *A journey towards regional economic integration: 1967–2017*. The ASEAN Secretariat.

ASEAN Briefing. (2015). *ASEAN open skies policy to be implemented in 2015*. https://www.aseanbriefing.com/news/2015/01/02/asean-open-skies-policy-implemented-2015.html

Balsalobre-Lorente, D., Driha, O. M., Shahbaz, M., & Sinha, A. (2020). The effects of tourism and globalization over environmental degradation in developed countries. *Environmental Science and Pollution Research, 27*(7), 7130–7144.

Capistrano, R. C., & Notorio, P. A. (2020). A content analysis of the future of tourism through the presidential state of the nation address in the Philippines (1987–2019). *Journal of Tourism Futures, 7*(1), 131–146.

Centre for Aviation. (2019). *Southeast Asia Airline 2019 Outlook: In this fast growth market low costs are essential*. https://centreforaviation.com/analysis/airline-leader/southeast-asia-airline-2019-outlook-in-this-fast-growth-market-low-costs-are-essential-457920

Chen, J. W. (2006). Achieving supreme excellence: How China is using agreements with ASEAN to overcome obstacles to its leadership in Asian regional economic integration. *China Journal International Leadership, 7*(2), 655–674.

Cohen, E. (2012). Globalization, global crises and tourism. *Tourism Recreation Research, 37*(2), 103–111.

Heng, J. (2020, May 5). Singapore clocks 4th straight record year for tourism arrivals and receipts in 2019. *Business Times*. https://www.businesstimes.

com.sg/government-economy/singapore-clocks-4th-straight-record-year-for-tourism-arrivals-and-receipts-in

Huang, Z. (2017). *One belt one road: Your guide to understanding BRI, China's new silk road plan*. https://qz.com/983460/BRI-an-extremely-simple-guide-to-understanding-chinas-one-belt-one-road-forum-for-its-new-silk-road/

Jetin, B. (2018). 'One belt-one road initiative' and ASEAN connectivity: Synergy issues and potentialities. In *China's global rebalancing and the new silk road* (pp. 139–150). Springer.

Kostková, M. (2020). *Shared accommodation as a globalization trend in tourism*. SHS Web of Conferences, 74. https://doi.org/10.1051/shsconf/20207403002

Kyrylov, Y., Hranovska, V., Boiko, V., Kwilinski, A., & Boiko, L. (2020). International tourism development in the context of increasing globalization risks: On the example of Ukraine's integration into the global tourism industry. *Journal of Risk and Financial Management, 13*(12), 303. https://doi.org/10.3390/jrfm13120303

Lau, R. (2017). China's belt and road: What's in it for Malaysia? *The Borneo Post Online*. http://www.theborneopost.com/2017/09/03/chinas-belt-and-road-whats-in-it-for-malaysia/

Lee, E., & Vivarelli, M. (2006). The social impact of globalization in the developing countries. *International Labour Review, 145*(3), 167–184.

Li, T., Shi, H., Yang, Z., & Ren, Y. (2020). Does the belt and road initiative boost tourism economy? *Asia Pacific Journal of Tourism Research, 25*(3), 311–322.

Ministry of Tourism & Sports Thailand. (2017). *The second national tourism development plan (2017–2021)*. Office of Permanent Secretary Ministry of Tourism and Sports Thailand.

Mohamad, M. (2002). *Globalization and the new realities*. Pelanduk.

Saint Akadiri, S., Lasisi, T. T., Uzuner, G., & Akadiri, A. C. (2019). Examining the impact of globalization in the environmental Kuznets curve hypothesis: The case of tourist destination states. *Environmental Science and Pollution Research, 26*(12), 12605–12615.

Scott, N. (2007). *Impact assessment of the visit ASEAN campaign*. Concept paper, The University of Queensland.

Song, H., Li, G., & Cao, Z. (2018). Tourism and economic globalization: An emerging research agenda. *Journal of Travel Research, 57*(8), 999–1011.

Tawil, R. F., & Al Tamimi, A. M. (2013). Understanding Chinese tourists' travel motivations: Investigating the perceptions of Jordan held by Chinese tourists. *International Journal of Business and Social Science, 4*(17), 164–170.

The ASEAN Post. (2021). *Southeast Asia soars with low-cost airlines*. https://theaseanpost.com/article/southeast-asia-soars-low-cost-airlines

The Organization for Economic Co-operation and Development (OECD). (2020). *OECD tourism trends and policies 2020*. https://www.oecd-ilibrary.org/sites/6c4fc1cd-en/index.html?itemId=/content/component/6c4fc1cd-en

Tourism Malaysia. (2018). *About tourism Malaysia*. https://www.tourism.gov.my/about-us/about-tourism-malaysia

United Nations. (2004). *Meeting the challenges in an era of globalization by strengthening regional development cooperation*. United Nations Publication.

UNWTO (2011). *Tourism highlights 2011 edition*. http://mkt.unwto.org/sites/all/files/docpdf/unwtohighlights11enlr.pdf

World Economic Forum. (2013). *Nelson Mandela's 1999 Davos address*. https://www.weforum.org/agenda/2013/12/nelson-mandelas-address-to-davos-1999/

World Travel & Tourism Council. (2016). *Travel and tourism in ASEAN is growing faster than in other regions of the world*. https://asean.psu.ac.th/Data/topic/1/t000072.pdf

Zhou, Y., Lin, X., Zhong, Y., & Xie, W. (2019). Contract selection for a multi-service sharing platform with self-scheduling capacity. *Omega-International Journal of Management Science, 86*, 198–217.

13

Marketing Emerging Tourist Destinations During Crisis and Pandemics

Ishmael Mensah

Introduction

One noticeable feature of emerging economies is their susceptibility to crises, which then tends to adversely impact the tourism industry. A crisis refers to "a low-probability, high-consequence event that develops very rapidly and involves ambiguous situations with unknown causes and effects" (Robert et al., 2007, p. 109). A tourism crisis can be described as:

> an event which causes disruption in the normal functioning of tourism-related businesses, damages all the reputation of the safety, attractiveness, and comfort of a touristic destination by negatively affecting the perceptions of visitors, as a result, negatively affects tourism economy and causes reduced tourist arrival and spending, delay doing business in the tourism sector. (Glaesser, 2006, p. 6)

I. Mensah (✉)
University of Cape Coast, Cape Coast, Ghana
e-mail: ikmensah@ucc.edu.gh

I. Mensah et al. (eds.), *Marketing Tourist Destinations in Emerging Economies*, Palgrave Studies of Marketing in Emerging Economies,
https://doi.org/10.1007/978-3-030-83711-2_13

Though a crisis is usually an unexpected adverse event, it could also be self-inflicted (Ritchie, 2009). Crises come in different forms, including financial crisis, energy crisis, political issues, epidemics/pandemics, and natural disasters. The tourism industry has long been recognised as one of the industries which are most vulnerable to crises or disasters (Hall, 2010). Crises tend to have adverse effects on tourist destinations, as they can disrupt the operations of tourism businesses, affect people physically and mentally, and endanger the viability of businesses that can no longer cope with the situation using normal managerial procedures (Okumus & Karamustafa, 2005). The impacts of crises on destinations in emerging economies are even more severe since their tourism industries are characterised by small to medium tourism enterprises. Studies have shown that small tourism businesses have neither the time nor resources needed to plan for crises (Cioccio & Michael, 2007). Thus, governments of emerging economies and destination management organisations (DMOs) have had to deal with the debilitating effects of these crises. This calls for a carefully planned and executed crisis response plan aimed at ensuring the recovery of the tourism industry and bolstering visitor confidence in destinations.

Forms of Crises

Throughout history, the tourism industry has experienced many crises and disasters, including the Bali bombing in 2002, the spread of Severe Acute Respiratory Syndrome (SARS) in the Asia-Pacific region in 2003, the Indian Ocean tsunami in 2004, the Thai political conflict during 2013–2014, the earthquake in Nepal in 2015, and the Ebola virus epidemic in West Africa between 2014 and 2016. Crises differ in terms of their occurrence and impact. There have been several attempts to classify such crises. Glaesser (2006) classifies them broadly into natural and man-made. Natural crises are natural disasters such as hurricanes, volcanic eruptions, forest fires, and earthquakes, while man-made crises have an element of human intent or error, such as terrorist attacks, arson, and political unrests. Crises have also been categorised into ecological, financial, regional, and global (Okumus & Karamustafa, 2005). Generally, crises that adversely impact the tourist destination could be classified into financial crises, political crises, energy crises, epidemics/pandemics, conflicts, and natural disasters.

Financial Crises: There have been several financial crises throughout history, including the great depression of 1929–1939, the Asian financial crisis of 1997, and the global financial crisis of 2008–2009. This usually occurs when some significant financial assets such as oil, real estate, stocks, loans, or investments unexpectedly lose a substantial amount of their nominal value. For instance, the global financial crisis of 2008–2009 was the result of a slump in the value of derivatives that were being used for hedge fund trading by banks, and as a result, the banks stopped lending to each other. Financial crises normally result in an economic recession. Recession affects the availability of credit and general economic activities. As a result, businesses cut costs by laying off workers, and hence unemployment rates increase. Recession negatively impacts travel and the competitiveness of tourist destinations (Perles-Ribes et al., 2016). However, Perles Ribes and Ramón Rodríguez (2013) found that in emerging destinations with a growing natural trend on tourism demand, the effects of economic crisis are softer and limited.

Energy Crises: Energy crises result from a significant challenge in the supply of energy resources to destinations. Most of the energy crisis has been occasioned by declining oil stocks because fossil fuel, which is a non-renewable energy source, is the main source of global energy supply. While demand for oil is on the ascendancy, its supply has been relatively stagnant. Between 2012 and 2016, Ghana, for instance, experienced a protracted electricity crisis dubbed *dumsor*, which significantly affected businesses including hotels and restaurants. Many hotels, for instance, had to invest in generators, thereby increasing operational costs.

Epidemics/Pandemics: A pandemic is a widespread event involving an infectious disease that spreads across international frontiers, while an epidemic is the outbreak of a disease in a particular community within a particular time. Thus, an epidemic becomes a pandemic when it spreads to multiple countries or regions of the world. These include diseases like influenza, yellow fever, cholera, smallpox, and polio. Throughout history, there have been periodic outbreaks of epidemics and pandemics, which rendered travel unsafe. These include the Black Death (1346–1353),

Spanish flu (1918–1920), SARS (2002–2004), H1N1 Swine flu (2009–2010), Middle East Respiratory Syndrome (MERS) (2012), and Ebola Virus (2014–2016). The Spanish flu, for instance, restricted travel for four months and killed 21 million people during that short period.

During the twentieth century, there were three major pandemics, namely the Spanish flu (1918–1920), the Asian flu (H2N2) of 1957, and the Hong Kong flu of 1968. The outbreak of pandemics has increased in the twenty-first century. The major pandemics during the period were SARS in 2002, "Bird flu" in 2009, MERS in 2012, and Ebola, which peaked in 2013–2014, and COVID-19 from 2019 to date. The COVID-19 pandemic erupted in December 2019 in the city of Wuhan, in the Hubei Province, in China and spread rapidly to other parts of the world through human-to-human transmission.

However, in terms of the speed of spread and impact, COVID-19 is the most serious crisis the tourism industry has ever encountered. The impact of COVID-19 on the tourism industry is estimated to be seven times more than the impact of the 2008 financial crises (Cetin, 2020).

In less than six months, COVID-19 had infected 4,716,965 people and killed 315,248 people as recorded by the dashboard of the Center for Systems Science and Engineering at Johns Hopkins University (Dong et al., 2020). As on 12 March 2021, the number of people infected had increased to 118,650,470, while the number of deaths had reached 2,631,694 (Dong et al., 2020). Çakmaklı et al. (2020) predict that the COVID-19 shock may soon result in the biggest emerging market crisis in modern times. It will result in a collapse in domestic and external demand, capital outflows, increased cost of external borrowing, a commodity bust, and depreciation in currencies.

Political Unrests and Conflicts: Political unrests include coup d'état, anti-government demonstrations, general strikes, riots, and other uprisings. Other political risks include wars, revolutions, acts of terrorism, and other forms of violence and crime. Such political upheavals such as the 1973 Arab-Israeli War (*Yom Kippur War*), the civil war in the former Yugoslavia in the early 1990s, political unrest in Fiji in 2006, Arab Springs between 2010 and 2011, and Myanmar coup d'état in 2021 have had significant impacts on tourism in and around these countries. The Arab Spring in the Middle East, for instance, caused a sharp decline in tourism in the region (Masetti et al., 2013). Tourism destinations and tourists have usually been "soft targets" for terrorist activities (Fuchs & Pizam, 2011). A case in point is the Luxor massacre in Egypt in 1997, when extremists ambushed and killed 62 tourists. Egypt has suffered a series of terrorist attacks since the 1990s, some of which were directed at foreign tourists. Terrorists, by attacking tourist destinations and tourists indiscriminately, seek to attract attention to their causes in a most effective manner (Sönmez et al., 1999). Generally, political upheavals which are rampant in the developing world tend to heighten the risk profile of tourist destinations, rendering them unattractive for travel.

Natural Disasters: These are climate-oriented and geophysical disasters. They include earthquakes, hurricanes, tornadoes, floods, tsunamis, storms, bush fires, droughts, and heatwaves and volcanic eruptions, which occur as a result of natural processes. These include the 2008 Cyclone Nargis in Myanmar, the 2010 Haiti earthquake, the 2010 Mt. Eyjafjallajokull eruption in Iceland, the 2011 Tsunami in Japan, and the 2015 earthquake in Kathmandu, Nepal. Asia is considered the most disaster-prone region in the world (Konrad, 2013). These natural disasters had a significant impact on tourism activities and tourist destinations. According to Jin et al. (2019), the repercussions of such disasters are likely to affect tourism directly at a destination country and are indirect consequences for travel to and from the affected region. For instance, water shortage in Cape Town, South Africa, resulted in a reduction in tourism and a notable loss in income for local businesses (Rosselló et al., 2020).

Accidents: Accidents are unexpected events that result in damage, injury, or loss of lives. These include road accidents, aeroplane crashes, and the sinking of ships. It also includes the accidental release of harmful chemicals and biological agents into the air, land, or water. These include accidents such as the Bhopal gas disaster in India in 1984, the Exxon Valdez oil spill into Alaska's Prince William Sound in 1989, and nuclear accidents like the Chernobyl nuclear disaster in 1986 and the disappearance of Malaysia Airlines Flight 370 in 2014. Accidents could also be self-inflicted such as a tourist sustaining injury as a result of engaging in dangerous sports such as skydiving or bungee jumping. Such incidents also render destinations unsafe for travel.

Pandemics and Tourism

There has always been a link between travel and pandemics. Travel fuels the spread of pandemics, and the spread of pandemics becomes a disincentive for travel. The relationship between tourism and pandemics is rather ironic, in the sense that while travellers may become vectors for the spread of the disease, travel and tourism may eventually become the main victim (Wilder-Smith, 2006).

Throughout history, travel has contributed to the spread of pandemics, with transportation acting as a vector for the transmission of viruses. The Spanish flu, for instance, was spread rapidly by soldiers who travelled back home in ships after the First World War. Also, the international spread of SARS in 2003 was through travel by a physician from Guangdong to Hong Kong. While in Hong Kong, he stayed in a hotel and infected other guests from Hong Kong, China, Vietnam, Singapore, and Canada. Those who were infected subsequently travelled to their respective countries and infected other people, and this sparked off an international spread of SARS (Abdullah et al., 2004; Hung et al., 2018). In the case of COVID-19, Gössling et al. (2020) indicated that it was the global air transport that carried the virus to all continents, to the extent that by mid-March, it had spread to 146 countries. Additionally, Monterrubio (2010) has suggested that it is travel that determines the actual routes that a disease takes, when the disease reaches different parts of the world, and how it is transmitted.

In view of this, the travel and tourism industry has often been targeted for stemming the tide of the spread of diseases. Governments have often been compelled to impose travel bans to manage transmission risks (Gössling et al. 2020). To reduce the spread of the COVID-19 pandemic, to "flatten the curve", almost every country has imposed and enforced border closures, travel restrictions, quarantine, and lockdowns, as well as cancellation of flights and events. In the same vein, the World Health Organization (WHO) has been at the forefront of issuing travel advisories to help control the spread of pandemics. For instance, during the outbreak of SARS in 2003, WHO issued travel advisory recommendations from 2 April to 23 May 2003, asking travellers to consider postponing all non-essential travels to high-risk areas including Hong Kong, Beijing, Toronto, and the Guangdong Province in Mainland China (WTTC, 2003).

Impacts of Crises and Pandemics on Tourism in Emerging Economies

Crises such as political unrests, earthquakes, and pandemics have had a devastating toll on the tourism industries of emerging economies. Crises events generally render destinations unsafe for travel and tourism activities. Apart from the social and economic effects of such crises, they also tend to have devastating effects on the tourism industry (Mason et al., 2005). The 2014–2015 Ebola virus outbreak, for instance, had a negative impact on West African countries, with Sierra Leone suffering the most (Kongoley, 2015). Also, in 2003, the outbreak of the SARS virus created a huge negative impact on the growth of tourism in China (Zeng et al., 2005).

Because crises create an atmosphere of insecurity at a destination, they render the destination unattractive to tourists. This is because the perception of safety and security at a destination is a major determinant of travel behaviour (Beirman, 2003). Perceived insecurity at a destination creates fear among tourists, who are generally risk-averse. Thus, the real or perceived threat to the health, safety, or security of tourists is likely to influence their decision on travel to a particular destination (Lepp & Gibson, 2003). In Egypt, the terrorist attacks which began in 1992 resulted in a 43% drop in tourism receipts (Wahab, 1996). A study by Zheng et al. (2021) found that the severity of and susceptibility to threat after the outbreak of a pandemic can cause travel fear, which then leads to protection motivation and then protective travel behaviours. The study further revealed that the fear of travel can evoke different coping strategies, which increases people's psychological resilience and adoption of cautious travel behaviours. Among the cautious travel behaviours is the choice of alternative destinations. According to Prideaux et al. (2003, p. 475), "tourism flows are subject to disruption by a range of events that may occur in the destination itself, in competing destinations, origin markets, or they may be remote from either". Crises events usually result in the diversion of tourism flows away from not only the destination with the crisis but also to neighbouring regions or countries (Cavlek, 2002).

The real or perceived dangers of travel to destinations that have been hit by crises result in a significant decline in tourist demand (Lepp & Gibson, 2003; Mawby, 2000). Usually, the natural response to crises is the cancellation of scheduled trips. Though crises, in general, have resulted in the cancellation of planned trips, it appears the situation is more precarious with pandemics. During pandemics, tourists cancel their travels, avoiding places and people associated with the pandemic (Nicholl, 2006). During the outbreak of SARS in 2003, most of the scheduled travel by overseas tourist groups to Beijing between April and May 2003 were cancelled, and this resulted in arrivals by foreign tourists dropping by 59.9% compared with the same period in 2002 (People's Daily, 2003, June). Also, in Hong Kong, it resulted in massive cancellations of flights, package tours, conventions, business trips, and hotel reservations (Lo et al., 2006). Mexico had a similar experience during the outbreak of swine flu (N1H1) when international travel and tourism companies cancelled their trips to Mexico. Thomas Cook, UK, cancelled its holidays to Cancun, Mexico, up to 5 May 2009, despite WHO's travel advisory to avoid travel restrictions. Most countries advised against travelling to Mexico, and many repatriated their citizens (Monterrubio, 2010). COVID-19 has had a more devastating impact on travel and tourism in emerging economies. Due to the COVID-19 pandemic, ticket refunds increased by 75% between February and March 2020, compared to the same period in 2019 (International Air Transport Association [IATA], 2020).

The fear of travelling to destinations devastated by crises has resulted in significant declines in tourist arrivals in emerging economies. During the height of terrorist attacks in the 1980s and the Persian Gulf War in 1991, there were significant declines in visitations to the Middle East (Sönmez et al., 1999). Several crises, including the SARS epidemic of 2002, the Tsunami of 2004, the global economic crisis of 2007 and 2010, riots and political uncertainty in 2010, the floods of Thailand and northern Bangkok in 2011, and anti-government demonstrations and coup d'état between 2013 and 2014, resulted in a reduction in inbound tourism and drove the Thai tourism industry on a downward spiral between 2002 and 2014 (Chetthamrongchai, 2017). In West Africa, the outbreak of Ebola in 2014 also generated fear of travel to the region, in particular, and Africa as a whole, although 99% of the cases were recorded in Sierra Leone, Guinea, and Liberia. This resulted in a 7.7% decline in arrivals in other Ebola-free countries. Even after 35 months of the Ebola outbreak, international arrivals to West African countries were more than 50% short of the pre-epidemic levels (World Travel & Tourism Council [WTTC], 2018). Countries as far away as Kenya, over 3000 miles from the outbreak, reported a significant loss in arrivals during the period, citing travel fears over Ebola among other factors (WTTC, 2018).

The decline in tourist arrivals resulting from crises also leads to shortfalls in tourism revenue. Soemodinoto et al. (2001) reported that large-scale riots on the Gili Islands of Indonesia from October 1999 to April 2000 resulted in a significant decline in tourist arrivals by 6860 and a concomitant loss of US$329,280 to US$658,560, compared to the same period the previous year. Also, a small country like Brunei lost almost 30,000 tourists, which accounted for 15% of the total number of tourists and B$15 million, due to the swine flu during the first 12 months after the outbreak of the flu (Haque & Haque, 2018). The swine flu was also estimated to have cost Mexico nearly a million overseas visitors and around US$2.8 billion over five months (Russy & Smith, 2013). In Singapore, one of the most popular tourist destinations in Southeast Asia, the outbreak of SARS caused visitor arrivals to drop significantly from April 2003 such that by May, it was 70% lower than the previous year's figures (Singapore Tourist Board [STB], 2003a). Similarly, in Sierra Leone, the outbreak of Ebola saw tourist arrivals down by 50% from 2013 to 2014 (WTTC, 2018).

The impact of the COVID-19 pandemic on tourist arrivals and tourism revenue on the global economy, in general, and emerging economies, in particular, is unprecedented.

The UNWTO (2020a) has projected a 20–30% decline in 2020 international arrivals that would translate into losses in tourism receipts of US$300–450 billion. According to the UNWTO, this is even worse than the impact of SARS in 2003 and the global economic crisis in 2009

(UNWTO, 2020b). The WTTC provided an even higher estimate of a loss of up to US$2.1 trillion in 2020. It appears no country has been spared of the scourge of the raging COVID-19 pandemic. The COVID-19 pandemic has not spared any emerging economy and indeed any advanced economy. It was estimated that it would cost the Chinese tourism industry approximately 900 billion yuan (US$128 billion) in the first two months of 2020 compared with the previous year (Leng, 2020, February 22). Also, Vietnam, which received approximately 1.45 million Chinese visitors in the first quarter of 2019, experienced a drop in the arrivals of 644,000 in January 2020. It was estimated that this would cost the country US$5 billion by the second quarter of 2020 (Vietnam Times, 2020).

In times of crises, as cancellation of trips, bookings and events, as well as border closures and lockdowns, ensue and tourist arrivals decline, tourism businesses at the destination suffer the most. Hotels, airlines, restaurants, convention centres, attractions, travel agencies, and tour operators experience significant declines in business, revenues, and cash flow. According to Tounta (2020), history has taught us that epidemics and pandemics have an instant effect on tourism businesses, including hotels, restaurants, airlines, and travel agencies due to international travel restrictions, media coverage, and government measures. The situation is even more precarious for emerging economies since such destinations are usually characterised by smaller businesses like budget hotels, souvenir shops, and travel agencies that are not resilient enough to withstand the shocks associated with these crises and may have to close down.

Airlines are usually the first to suffer such declines due to flight cancellations and the accompanying demands for refunds. The IATA (2020) predicted that the COVID-19 pandemic would reduce global airline revenue by US$29.3 billion in 2020 due to a contraction in global air demand. By 11 March 2020, African airlines had lost US$ 4.4 billion in revenue due to COVID19, with Ethiopian Airlines, one of the largest airlines on the continent, reporting a loss of US$190 million (IATA, 2020).

Hotels are inextricably linked to travel and, therefore, during periods of declines in travel, the hotel business is seriously hampered, as they suffer declines in occupancy and revenue.

In 2003, average occupancy rates across the hotel industry fell by almost 30% between May and June in SARS-affected destinations

(WTTC, 2003). Also, in Hong Kong, the average hotel occupancy level among member hotels of the Hong Kong Hotels Association (HKHA) was about 20% in April 2003 compared to 85% during the same period in the previous year (Hong Kong Hotels Association, 2004). The situation in Singapore was no different, as the average occupancy rate during the second quarter of 2003 for hotels, in general, was 21%, compared with 74.5% for the previous year, while average room rates reduced by 18.8% (STB, 2003a). Similarly, during the swine flu pandemic in Mexico, average hotel occupancy rates declined significantly to between 4% and 30%, compared to the 80% recorded at the same time in the previous year (Monterrubio, 2010). The picture is even bleaker in a COVID-19 era. In China, after the outbreak of the pandemic, the hotel industry witnessed a 71% year-on-year decline in occupancy from 23 January to 26 January (Baker, 2020).

Loss of revenue, cash flow problems, and a general decline in business activities compel tourism businesses to adopt remedial measures, which often include lay-offs and job cuts. The resultant cancellations and reduced occupancy mean that tourism businesses operate under their full capacity during crises. These businesses, therefore, cannot support their full human resource capacity. Many employees are, therefore, laid off or placed on furlough during crises. In 2003, over three million employees in the tourism industry lost their jobs after the outbreak of SARS, causing an economic loss of over US$20 billion to East Asian economies (WTTC, 2003). WTTC (2020) has warned that the COVID-19 pandemic could slash 50 million jobs globally in the travel and tourism industry, and Asia will be the worst affected.

Crises and Destination Image

Destination image has been described as the tourist's general perception of a destination (Tasci et al., 2007). It refers to an individual's beliefs, thoughts, and impressions about a destination (Crompton, 1979). It is the impressions that people form about a destination.

Destination image plays a vital role in people's travel decision-making, including destination choice behaviour (Govers & Go, 2003; O'Leary &

Deegan, 2005). Thus, the decision to visit or not to visit a particular destination could be influenced by the image that a tourist holds about a place. If people have a negative perception about a destination, they are unlikely to visit such a destination. This is why a crisis-ravaged destination is usually not attractive to tourists. Singapore's long-established reputation as a safe and secure destination was seriously dented during the outbreak of SARS in the region (Henderson, 2003). Many countries in the Middle East have a negative image as a result of prolonged political upheavals, and this is a barrier to tourism development (Avraham, 2016). Political violence tends to affect tourists' perceptions of a destination and travel behaviour, as exemplified by countries like Egypt, Israel, Northern Ireland, and Peru (Sönmez et al., 1999). Sub-Saharan African (SSA) countries especially have suffered from prolonged negative place image, stereotypes, and generalisations (Ketter & Avraham, 2010; Muhwezi et al., 2016). This is mainly a result of media reportage on political unrests, violent conflicts, underdevelopment, serious corruption, severe droughts, and extreme poverty (Boniface et al., 2012). The resulting negative images of SSA countries have served as significant obstacles to attracting many international tourists (Muhwezi et al., 2016).

Crises involving safety and security incidents are of interest to the media, as they are regarded as important news generators (Pennington-Gray & Pizam, 2011). However, the level of sensationalism and selectivity with which the media reports such events usually contributes to the spread of fear and panic. The media influence people's perceptions of a destination by providing them with very vivid and explicit information and analyses of such incidents (Tasci & Gartner, 2007). Usually, when there is a crisis such as war, insurgency, and epidemics, it is the media that rapidly draw people's attention to it, thereby shifting tourists' travel intentions to alternative destinations (Scott et al., 2008). Sometimes, the media report the crises with bias and exaggeration. For instance, the Indian Ocean tsunami of 2004 devastated only a small part of the island of Phuket, but the media reported that the island had been destroyed (Gurtner, 2007). Sometimes, the alarmist and sensationalist nature of the news reports by the international media worsens the effect of the crisis on the tourism industry (McKercher, as cited in Mason et al., 2005). Avraham and Ketter (2008) have reported on how international media reportage on crises including terrorist attacks, wars, and political instability destroyed many tourist industries. Wen et al. (2020) reported that China attracted a lot of bad press over the outbreak of the COVID-19 pandemic due to global concerns about the virus, which originated from Wuhan, China. Some media channels initially inappropriately labelled the disease "Chinese virus pandemonium" before the official name was confirmed. The *Wall Street Journal* in February 2020 published an article entitled "China is the Real Sick Man of Asia". Also, a local French newspaper *Le Courier Picard* used the headlines "Alerte Jaune" (Yellow alert) and "Le péril Jaune?" (Yellow peril?), with an accompanying image of a Chinese woman wearing a protective mask. These publications, which attracted a backlash from the Chinese at home and abroad, impacted adversely on the image of China as a tourist destination.

The negative publicity that media reports on crises engender often becomes associated with a destination even after the occurrence of the crisis and lasts until there is full recovery and the destination bounces back to normal. According to Sönmez et al. (1994), this period can represent a tourism crisis for a destination, whereby the normal operation and conduct of tourism-related businesses would be hampered. This will affect the destination's image and cause a decline in tourist arrivals and expenditures.

Crisis Communication

However, destinations engulfed in crisis could also employ the media to correct those negative images and stereotypes associated with the destination as a result of the crisis. This calls for an effective crisis communication strategy. Armstrong and Ritchie (2008, p. 187) emphasised that crisis communication is an important role of destination marketing organisations (DMOs), as it helps to reduce the negativity associated with the crisis, bring both private and public sectors together, and implement consistent messages and recovery marketing methodologies. Ritchie (2004) has also emphasised the importance of crisis communication to the recovery of a destination after a crisis, indicating that a comprehensive communication strategy ensures that DMOs are proactive and effective rather than reactive and ad hoc. Emerging economies are usually reactive in crisis management, and this could be attributed to that lack of effective crisis communication strategies when crises ensue. For instance, Pforr and Hosie (2008) have lamented that many Southeast Asian governments have demonstrated a preference for reactive crisis management and have developed specific responses or practical guidelines on how to respond to a crisis.

According to Young and Montgomery (1998, p. 4), "a crisis has the potential to be detrimental to the marketability of any tourism destination, particularly if it is dramatized and distorted through rumours and the media". However, when a communication plan is well integrated into the overall crisis management plan of a destination, it can result in rapid depletion of negative images and ensure a quicker recovery (Faulkner, 2001). Crisis communication usually has several goals including correcting wrong perceptions about the extent of the crisis, restoring visitor confidence in the destination, reducing visitors' perceptions of risk, and repairing the destination's image (Walters & Mair, 2013). There has been a considerable debate on the communication strategies suitable for a crisis.

Benoit (2015) classifies these strategies broadly into denial (simple denial, shift the blame), evading responsibility (provocation, defeasibility, accident, good intentions), reducing offensiveness (bolstering, minimisation, differentiation, transcendence, attack accuser, compensation), corrective action, and mortification. Ultimately, a crisis communication strategy seeks to address the negative images associated with a destination resulting from the crisis and rekindle visitor interest in the destination. Avraham and Ketter (2017) proposed the multi-step model for altering the post-crisis image of a destination. These are source strategies, audience strategies, and message strategies. *Source Strategies* focus mainly on the marketers' efforts to influence or replace the source that they believe underlies the destination's image, such as the international media that project the negative images. The strategy could be to organise familiarisation tours for foreign journalists. Following the Izmit earthquake, Turkish tourism offices, in collaboration with the Turkish Ministry of Tourism, Turkish Airlines, and major tour operators servicing Turkey, invited the media and travel agents to see the extent of the damage. They visited the main tourist areas to reinforce the message that the areas were largely untouched by the earthquake (Beirman, 2003).

Audience Strategies are geared towards shaping the audience's values, perceptions, and dreams. The marketers try to show the commonly held values by their country and the foreign audience. A typical example was the campaign sponsored by Saudi Arabia to restore its image in the United States after the 9/11 crisis by emphasising the country's special and unique ties to America over the years (Avraham, 2013).

Message Strategies seek to contradict the negative messages, stereotypes, perceptions, and generalisations about the destination that the crisis has generated. In 2014, the Kenya Tourism Board launched a social media campaign under the theme, "#WhyILoveKenya", to recover its image, which had been badly bruised by a series of terrorist attacks. People were encouraged to tell their stories about Kenya by uploading stories, pictures, and videos of their peaceful travel to Kenya, using the hashtag, "#WhyILoveKenya" (Avraham, 2017).

To execute these strategies, destinations could choose several techniques including advertising, promotions, and public relations to transmit the message through different media such as television, radio, press, and Internet as exemplified by Egypt after the political crisis of 2011–2014. To restore a positive image of the destination, a variety of creative marketing communications activities were undertaken, including advertising campaigns, press reports and public relations activities (Avraham, 2016).

In terms of the contents of messages, Walters and Mair (2013) identified a number of major themes that have been used in the past as marketing messages to encourage tourists following a disaster. These include open and ready for business, solidarity messages, community readiness, restoration of confidence, curiosity enhancement, as well as visitor or celebrity testimonials.

According to them, 'open and ready for business' is the most common type of message used by DMOs eager to see the destination bounce back as soon as possible. This is to reassure visitors that the destination has overcome the crisis and is back to business. This type of message was used in the Maldives after the Boxing Day Tsunami (Carlsen & Hughes, 2008).

Solidarity messages are used to evoke an emotional reaction among potential visitors by courting their sympathy and to counteract any feeling of guilt on the part of visitors who might think they are getting in the way of recovery efforts. The Maldives employed this strategy with the message "by visiting the Maldives, you are helping us".

'Community readiness' is a type of massage that seeks to reassure potential visitors that the community is ready to receive them after the disaster. For instance, "We are ready to welcome you" was used in Sri Lanka following the Boxing Day Tsunami (Robinson & Jarvie, 2008). During crises, tourists may lose confidence in the affected destination due to negative media reportage. 'Restoration of confidence' is a message that seeks to reassure visitors that the effect of the disaster is not as serious as they have been made to believe. This message strategy was used to promote Canberra after the bushfires in 2004, using the message, "our heart's still going strong" (Armstrong & Ritchie, 2008).

Curiosity enhancement messages seek to generate traveller curiosity to revisit the destination by telling a story that focuses on the positive side of the destination. Studies have shown that curiosity is one of the main motivations for visiting a destination after a crisis (Rittichainuwat, 2008).

Destinations have also employed visitor testimonials or celebrity endorsement testimonials in crisis communication. Guest testimonials have been seen as particularly credible by potential tourists (Carlsen & Hughes, 2008). An example is the "Vive Mexico" campaign in 2009 to reignite tourism following the H1N1 pandemic. The Mexican government used endorsement testimonials from celebrities such as golfer Lorena Ochoa, singers Alejandro Fernandez and Placido Domingo, actor Diego Luna, and soccer player Rafael Marquez.

Crisis Management Strategies

The volatility and susceptibility of the tourism industry to crises underscore the need for strategies to deal with and overcome the impacts of crises in the industry (Gurtner, 2005; Pforr & Hosie, 2008). Crisis management is also imperative for the tourism industry because it is one industry that is often impacted adversely by external political, economic, social, and technological factors, which are often beyond the immediate control of the industry (Ritchie, 2004). According to Glaesser (2006), crisis management is composed of strategies, processes, and measures that are planned and put into force to prevent and cope with a crisis. It involves a strategic approach to planning for and managing crises.

Generally, crisis management involves three stages: pre-crisis, actual crisis, and post-crisis (Mair et al., 2016; Ritchie, 2009). The pre-crisis stage involves prevention, planning, and preparatory activities before the crisis. To ensure that destinations are adequately prepared for crises, several activities are undertaken, including risk assessment and management, creation of emergency response plans, and establishment of a crisis management team. The actual crisis involves recovery efforts geared towards minimising the effects of the crisis on the destination and its stakeholders. At this stage, the destination is mainly preoccupied with bringing the crisis to an end and limiting damage to tourists, residents, employees, businesses, and infrastructure. At the post-crisis stage, the crisis would have been over and businesses bounce back to normalcy or even get better. This stage also requires post-crisis recovery efforts to bring the destination back to normal (Mair et al., 2016). Also, at this stage, valuable lessons are learnt from the crisis to enable the destination to manage future crises.

At the heart of crisis management is recovery. According to Pennington-Gray and Pizam (2011), the purpose of the recovery function is to return systems to normal levels. Recovery ensures that the destination is restored to pre-crisis levels. Recovery strategies include damage assessment, disaster relief, reconstruction, crisis counselling, and government support in the form of financial incentives like tax cuts, soft loans, and employment support. Typically, tourism businesses on their own cannot recover easily from a crisis. The role of the government is very important. For instance, regarding the 9/11 disaster, Blake and Sinclair (2003) suggested that it was government policies such as sector-specific target subsidies and tax reductions that helped to recover the tourism industry after the incident.

During the SARS epidemic, the Singaporean government provided a S$230 million relief package for the tourism industry, including property tax rebates and loans for small and medium enterprises that were facing financial difficulties (Singapore Tourist Board, 2003b).

Also, during the swine flu pandemic, the Calderon administration created a special fund to provide loans and other direct support as well as tax cuts to hotels, restaurants, airlines, recreational service companies and other small and medium-sized enterprises that depended on tourism in Mexico (Monterrubio, 2010).

Conclusion

Tourist destinations, including those in emerging economies, are not immune to crises. When those crises occur, tourism is seriously hampered. Many emerging tourist destinations have experienced several crises including political, economic, natural, and health-related crises. Emerging economies have particularly suffered from the impacts of crisis and pandemics because they are less resistant to the shocks that such disasters engender. The tourism industry, in particular, is very sensitive to crises. The sense of insecurity that accompanies such crises, coupled with negative reportage by the media, tends to affect travel behaviour and choice of destination. The effect is that destinations that have crises situations often suffer a decline in tourist arrivals, tourism revenues, and tourism business activities as a whole. To minimise the impacts of crises on emerging economies, governments and DMOs, in collaboration with other stakeholders, must embark on crisis communication and crisis management.

References

Abdullah, A. S. M., Thomas, G. N., McGhee, S. M., & Morisky, D. E. (2004). Impact of severe acute respiratory syndrome (SARS) on travel and population mobility: Implications for travel medicine practitioners. *Journal of Travel Medicine, 11*(2), 107–111.

Armstrong, E. K., & Ritchie, B. W. (2008). The heart recovery marketing campaign: Destination recovery after a major bushfire in Australia's national capital. *Journal of Travel & Tourism Marketing, 23*(2–4), 175–189.

Avraham, E. (2013). Battling stereotypes of terror and wars: Media strategies for marketing tourism to Middle Eastern countries. *American Behavioral Scientist, 57*(9), 1350–1367.

Avraham, E. (2016). Destination marketing and image repair during tourism crises: The case of Egypt. *Journal of Hospitality and Tourism Management, 28*, 41–48.

Avraham, E. (2017). Changing the conversation: How developing countries handle the international media during disasters, conflicts, and tourism crises. *Journal of Information Policy, 7*, 275–296.

Avraham, E., & Ketter, E. (2008). *Media strategies for marketing places in crises: Improving the image of cities, countries, and tourist destinations*. Butterworth Heinemann.

Avraham, E., & Ketter, E. (2017). Destination marketing during and following crises: Combating negative images in Asia. *Journal of Travel & Tourism Marketing, 34*(6), 709–718.

Baker, T. (2020). *Chinese hotels seeing steep declines from coronavirus*. http://hotelnewsnow.com/Articles/300132/Chinese-hotels-seeing-steepdeclines-from-coronavirus

Beirman, D. (2003). *Restoring tourism destinations in crisis*. Allen & Unwin.

Benoit, W. L. (2015). *Accounts, excuses and apologies: Image repair theory and research* (2nd ed.). State University of New York.

Blake, A., & Sinclair, M. T. (2003). Tourism crisis management: US response to September 11. *Annals of Tourism Research, 30*(4), 813–832.

Boniface, B., Cooper, C., & Cooper, R. (2012). *Worldwide destinations: The geography of travel and tourism*. Routledge.

Çakmaklı, C., Demiralp, S., Kalemli-Özcan, S., Yesiltas, S., & Yildirim, M. A. (2020). *COVID-19 and emerging markets: An epidemiological model with international production networks and capital flows* (No. w27191). National Bureau of Economic Research.

Carlsen, J. C., & Hughes, M. (2008). Tourism market recovery in the Maldives after the 2004 Indian Ocean tsunami. *Journal of Travel & Tourism Marketing, 23*(2–4), 139–149.

Cavlek, N. (2002). Tour operators and destination safety. *Annals of Tourism Research, 29*(2), 478–496.

Cetin, G. (2020). Impact of Covid-19 on tourism industry. *Journal of Tourismology*, Editorial Letter.

Chetthamrongchai, P. (2017). The influence of travel motivation, information sources and tourism crisis on tourists' destination image. *Journal of Tourism and Hospitality, 6*(2), 278.

Cioccio, L., & Michael, E. J. (2007). Hazard or disaster: Tourism management for the inevitable in Northeast Victoria. *Tourism Management, 28*(1), 1–11.

Crompton, J. L. (1979). An assessment of the image of Mexico as a vacation destination and the influence of geographical location upon that image. *Journal of Travel Research, 17*, 18–23.

Dong, E., Du, H., & Gardner, L. (2020). An interactive web-based dashboard to track COVID-19 in real time. *The Lancet. Infectious Diseases, 20*(5), 533–534.

Faulkner, B. (2001). Towards a framework for tourism disaster management. *Tourism Management, 22*(2), 135–147.

Fuchs, G., & Pizam, A. (2011). The importance of safety and security for tourism destinations. In Y. Wang & A. Pizam (Eds.), *Destination marketing and management: Theories and applications* (pp. 300–313). CAB International.

Glaesser, D. (2006). *Crisis management in the tourism industry*. Elsevier Butterworth-Heinemann.

Gössling, S., Scott, D., & Hall, C. M. (2020). Pandemics, tourism and global change: A rapid assessment of COVID-19. *Journal of Sustainable Tourism, 29*(1), 1–20.

Govers, R., & Go, J. (2003). Deconstructing Destination Image in the Information Age. *Information Technology and Tourism, 6*, 13–29.

Gurtner, Y. (2007). Crisis in Bali: Lessons in tourism recovery. In E. Laws, B. Prideaux, & K. Chon (Eds.), *Managing tourism crises* (pp. 81–97). CABI Publishing.

Gurtner, Y. K. (2005). Adversity and resilience: A case study of crisis management in a tourist-reliant destination. In P. Tremblay & A. Boyle (Eds.), *Sharing tourism knowledge, Proceedings of the 2005 CAUTHE Conference* (pp. 196–198). Charles Darwin University.

Hall, C. M. (2010). Crisis events in tourism: Subjects of crisis in tourism. *Current Issues in Tourism, 13*(5), 401–417.

Haque, T. H., & Haque, M. O. (2018). The swine flu and its impacts on tourism in Brunei. *Journal of Hospitality and Tourism Management, 36*, 92–101.

Henderson, J. C. (2003). Managing a health-related crisis: Severe acute respiratory syndrome (SARS) in Singapore. *Journal of Vacation Marketing, 10*(1), 67–78.

Hong Kong Hotels Association. (2004). *Hong Kong Hotels Association executive director's review (2003–2004).* http://www.hkha.org/internal/hkha/news011.htm

Hung, K. K., Mark, C. K., Yeung, M. P., Chan, E. Y., & Graham, C. A. (2018). The role of the hotel industry in the response to emerging epidemics: A case study of SARS in 2003 and H1N1 swine flu in 2009 in Hong Kong. *Globalization and Health, 14*(1), 117.

International Air Transport Association. (2020). *International Air Transport Association: Airlines Financial Monitor January–February 2020.* https://www.iata.org/en/iata-repository/publications/economic-reports/airlinesfinancial-monitor%2D%2D-feb-2020/

Jin, X., Qu, M., & Bao, J. (2019). Impact of crisis events on Chinese outbound tourist flow: A framework for post-events growth. *Tourism Management, 74*, 334–344.

Ketter, E., & Avraham, E. (2010). Online tourism marketing for Sub-Saharan African countries: Battling stereotypes of high risk, hostility and underdevelopment. *International Journal of Tourism Policy, 3*(4), 318–331.

Kongoley-MIH, P. S. (2015). The impact of Ebola on the tourism and hospitality industry in Sierra Leone. *International Journal of Scientific and Research Publications, 5*(12), 542–550.

Konrad, A. S. (2013). *Climate change in Asia: Natural disasters in Asia.* http://ejap.org/environmental-issues-in-asia/natural-disasters-asia.html

Leng, S. (2020, February 22). *Coronavirus: China's economy lost US$196 billion in January, February, says ex-IMF official. South China Morning Post.* https://www.scmp.com/economy/china-economy/article/3051909/covid-19-likely-slashus185-billion-chinas-economy-january

Lepp, A., & Gibson, H. (2003). Tourist roles, perceived risk and international tourism. *Annals of Tourism Research, 30*, 606–624.

Lo, A., Cheung, C., & Law, R. (2006). The survival of hotels during disaster: A case study of Hong Kong in 2003. *Asia Pacific Journal of Tourism Research, 11*(1), 65–80.

Mair, J., Ritchie, B. W., & Walters, G. (2016). Towards a research agenda for post-disaster and post-crisis recovery strategies for tourist destinations: A narrative review. *Current Issues in Tourism, 19*(1), 1–26.

Masetti, O., Korner, K., Forster, M., & Friedman, J. (2013). *Two years of Arab spring: Current issues, emerging markets.* Deutsche Bank.

Mason, P., Grabowski, P., & Du, W. (2005). Severe acute respiratory syndrome, tourism and the media. *International Journal of Tourism Research, 7*(1), 11–21.

Mawby, R. I. (2000). Tourists' perceptions of security: The risk fear paradox. *Tourism Economics, 6*, 109–121.

Monterrubio, J. C. (2010). Short-term economic impacts of influenza A (H1N1) and government reaction on the Mexican tourism industry: An analysis of the media. *International Journal of Tourism Policy, 3*(1), 1–15.

Muhwezi, D. K., Baum, T., & Nyakaana, J. B. (2016). Dealing with negative symbolism of destinations with difficult heritage: Analysis of Uganda's image. *Journal of Hospitality Management and Tourism, 7*(3), 33–42.

Nicholl, A. (2006). *Personal (non-pharmaceutical) protective measures for reducing transmission of influenza: ECDC interim recommendations*. https://www.ncbi.nlm.nih.gov/pubmed/17213533

O'Leary, S., & Deegan, J. (2005). Career progression of Irish tourism and hospitality management graduates. *International Journal of Contemporary Hospitality Management, 17*, 421–432.

Okumus, F., & Karamustafa, K. (2005). Impact of an economic crisis evidence from Turkey. *Annals of Tourism Research, 32*(4), 942–961.

Pennington-Gray, L., & Pizam, L. (2011). Destination crisis management. In Y. Wang & A. Pizam (Eds.), *Destination marketing and management: Theories and applications* (pp. 314–326). CABI.

People's Daily. (2003 June 1a). Beijing striving to revive tourism. *People's Daily*. http://english.peopledaily.com.cn/200306/01/eng20030601_117542.shtml

Perles Ribes, J., & Ramón Rodríguez, A. (2013). Economic cycles, asymmetric crises and tourism competitiveness: Emerging versus mature destinations. *Economics Bulletin, 33*(3), 2080–2097.

Perles-Ribes, J. F., Ramón-Rodríguez, A. B., Sevilla-Jiménez, M., & Rubia, A. (2016). The effects of economic crises on tourism success: An integrated model. *Tourism Economics, 22*(2), 417–447.

Pforr, C., & Hosie, P. J. (2008). Crisis management in tourism: Preparing for recovery. *Journal of Travel & Tourism Marketing, 23*(2–4), 249–264.

Prideaux, B., Laws, E., & Faulkner, B. (2003). Events in Indonesia: Exploring the limits to formal tourism trends forecasting methods in complex crisis situations. *Tourism Management, 24*(4), 475–487.

Ritchie, B. W. (2004). Chaos, crises and disasters: A strategic approach to crisis management in the tourism industry. *Tourism Management, 25*, 669–683.

Ritchie, B. W. (2009). *Crisis and disaster management for tourism: Aspects of tourism*. Channel View Publications.

Rittichainuwat, N. (2008). Responding to disaster: Thai and Scandinavian tourists' motivation to visit Phuket, Thailand. *Journal of Travel Research, 46*(4), 422–432.

Robert, K. H., Madsen, P., & Desai, V. (2007). Organizational sensemaking during crisis. In C. M. Pearson, C. Roux-Dufort, & J. A. Clair (Eds.), *International handbook of organizational crisis management* (pp. 107–122). Sage.

Robinson, L., & Jarvie, J. (2008). Post-disaster community tourism recovery: The tsunami and Arugam Bay, Sri Lanka. *Disasters, 32*(4), 631–645.

Rosselló, J., Becken, S., & Santana-Gallego, M. (2020). The effects of natural disasters on international tourism: A global analysis. *Tourism Management, 79*, 104080.

Russy, D., & Smith, R. (2013). The economic impact of H1N1 on Mexico's tourist and pork sectors. *Health Economics, 22*(7), 824–834.

Scott, N., Laws, E., & Prideaux, B. (2008). Tourism crises and marketing recovery strategies. *Journal of Travel & Tourism Marketing, 23*(2–4), 1–13.

Singapore Tourist Board. (2003a). *Fact sheets*. January–December, STB

Singapore Tourist Board. (2003b). Government unveils $230 million SARS relief package. Press Release, 24 April, Singapore Tourism Board.

Soemodinoto, A., Wong, P. P., & Saleh, M. (2001). Effect of prolonged political unrest on tourism. *Annals of Tourism Research, 28*(4), 1056–1060.

Sönmez, S. F., Apostolopoulos, Y., & Tarlow, P. (1999). Tourism in crisis: Managing the effects of terrorism. *Journal of Travel Research, 38*(1), 13–18.

Sönmez, S. F., Backman, S. J., & Allen, L. R. (1994). *Managing tourism crises: A guidebook*. Clemson University.

Tasci, A. D. A., & Gartner, C. W. (2007). Destination image and its functional relationships. *Journal of Travel Research, 45*, 413–425.

Tasci, A., Gartner, W., & Cavusgil, S. (2007). Conceptualization and operationalization of destination image. *Journal of Hospitality & Tourism Research, 31*(2), 194–223.

Tounta, P. (2020, March 03). Pandemic 2020: The impact on tourism and the shadowy points. *Travel Daily News*. https://www.traveldailynews.com/post/pandemic-2020-the-impact-on-tourism-and-the-shadowy-points

UNWTO. (2020a). *Tourism and covid-19*. https://www.unwto.org/tourism-covid-19-coronavirus

UNWTO. (2020b). International tourist arrivals could fall by 20–30% in 2020. https://www.unwto.org/news/international-tourism-arrivals-could-fall-in-2020

Vietnam Times. (2020). *Vietnam's tourism sector could lose US$ 5bln due to COVID 19.* https://vietnamtimes.org.vn/vietnamstourism-sector-could-lose-us-5bln-due-to-covid-19-17829.html

Wahab, S. (1996). Tourism and terrorism: Synthesis of the problem with emphasis on Egypt. In A. Pizam & Y. Mansfeld (Eds.), *Tourism, crime and international security issues* (pp. 175–186). John Wiley.

Walters, G., & Mair, J. (2013). Post disaster recovery marketing messages for tourism destinations. In M. McCabe (Ed.), *Routledge handbook of tourism marketing.* Routledge.

Wen, J., Aston, J., Liu, X., & Ying, T. (2020). Effects of misleading media coverage on public health crisis: A case of the 2019 novel coronavirus outbreak in China. *Anatolia, 31*(2), 331–336.

Wilder-Smith, A. (2006). The severe acute respiratory syndrome: Impact on travel and tourism. *Travel Medicine and Infectious Disease, 4*(2), 53–60.

World Travel & Tourism Council. (2003). *Special SARS analysis: Impact of travel and tourism.* World Travel and Tourism Council.

World Travel & Tourism Council. (2018). *Impact of the Ebola epidemic on travel and tourism.* https://www.wttc.org/-/media/files/reports/2018/impact-of-the-ebola-epidemic-on-travel-and-tourism.pdf

World Travels and Tourism Council. (2020). *Impact of pandemic on travel and tourism.* https://www.wttc.org/priorities/crisis-preparedness/pandemics/

Young, W. B., & Montgomery, R. J. (1998). Crisis management and its impact on destination marketing: A guide for convention and visitors bureaus. *Journal of Convention and Exhibition Management, 1*(1), 3–18.

Zeng, B., Carter, R. W., & De Lacy, T. (2005). Short-term perturbations and tourism effects: The case of SARS in China. *Current Issues in Tourism, 8*(4), 306–322.

Zheng, D., Luo, Q., & Ritchie, B. W. (2021). Afraid to travel after COVID-19? Self-protection, coping and resilience against pandemic 'travel fear'. *Tourism Management, 83*, 104261.

Index

I. Mensah et al. (eds.), *Marketing Tourist Destinations in Emerging Economies*, Palgrave Studies of Marketing in Emerging Economies,
https://doi.org/10.1007/978-3-030-83711-2

The manufacturer's authorised representative in the EU is Springer Nature Customer Service Centre GmbH, Europaplatz 3, 69115 Heidelberg, Germany. If you have any concerns regarding our products, please contact ProductSafety@springernature.com

Printed and bound by CPI Group (UK) Ltd, Croydon, CR0 4YY
30/09/2026
02225459-0014